Studies in Space Policy

Volume 33

Series Editor

European Space Policy Institute, Vienna, Austria

The use of outer space is of growing strategic and technological relevance. The development of robotic exploration to distant planets and bodies across the solar system, as well as pioneering human space exploration in earth orbit and of the moon, paved the way for ambitious long-term space exploration. Today, space exploration goes far beyond a merely technological endeavour, as its further development will have a tremendous social, cultural and economic impact. Space activities are entering an era in which contributions of the humanities — history, philosophy, anthropology —, the arts, and the social sciences — political science, economics, law — will become crucial for the future of space exploration. Space policy thus will gain in visibility and relevance. The series Studies in Space Policy shall become the European reference compilation edited by the leading institute in the field, the European Space Policy Institute. It will contain both monographs and collections dealing with their subjects in a transdisciplinary way.

The volumes of the series are single-blind peer-reviewed.

More information about this series at http://www.springer.com/series/8167

Annette Froehlich
Editor

Outer Space and Cyber Space

Similarities, Interrelations and Legal Perspectives

Editor
Annette Froehlich
European Space Policy Institute
Vienna, Austria

ISSN 1868-5307 ISSN 1868-5315 (electronic)
Studies in Space Policy
ISBN 978-3-030-80025-3 ISBN 978-3-030-80023-9 (eBook)
https://doi.org/10.1007/978-3-030-80023-9

This Springer imprint is published by the registered company Springer Nature Switzerland AG
The registered company address is: Gewerbestrasse 11, 6330 Cham, Switzerland

Preface

Under the topic *Outer Space and Cyber Space: Similarities, Interrelations and Legal Perspectives*, the European Space Policy Institute (ESPI) and the German Aerospace Center (DLR) invited students and young professionals worldwide to submit an article in order to express their views on this very timely topic. This follow-up initiative to a previous worldwide space law essay calls for the analyses of a broad range of relevant aspects as the outer space and cyber space domain do not only present analogies but are also strongly interrelated. This may occur on various levels by technologies but also in regard to juridical approaches, each nevertheless keeping its particularities.

Since modern societies rely increasingly on space applications that depend on cyber space, it is important to investigate how cyber space and outer space are connected by their common challenges. Furthermore, this book discusses not only questions around their jurisdictions, but also whether the private space industry can escape jurisdiction by dematerializing the space resource commercial processes and assets thanks to cyber technology. In addition, space and cyber space policies are analyzed especially in view of cyber threats to space communications. Even the question of an extraterrestrial citizenship in outer space and cyber space may raise new views. Finally, the interdependence between space and cyber space also has an important role to play in the context of increasing militarization and emerging weaponization of outer space. Therefore, this publication invites questioning the similarities and interrelations between outer space and cyber space in the same way as it intends to strengthen them.

May 2021

Dr. Annette Froehlich
European Space Policy Institute (ESPI)
Seconded by German Aerospace Center (DLR)
Vienna, Austria

Contents

Chapter 1
Jurisdiction Over the Realms Unlocked by Technology: Outer Space and Cyberspace

Alexandros Eleftherios Farsaris

Abstract In light of the discussions concerning the potential international cooperation in the form of a United Nations treaty on cyberspace, space law can offer an interesting example for comparison. Being both cyberspace and outer space areas where access was made possible only through the development of technology, the legislative path towards a jurisdictional regime can find similarities. Whilst cyberspace is inherently different from any other space regulated under international law, the traditional legal tools that were used in the creation of the jurisdictional regime in outer space, can definitely be of help when building the legal base for cyberspace. Nevertheless, whereas space law provisions can offer a useful tool for the understanding of the applicability of jurisdictional rules, the very nature of cyberspace requires a legal regime that cannot be built solely on parallelisms and established precedents, but requires the creation of a new legal paradigm.

1.1 Introduction

Starting with the launch of the first man-made object in outer space in 1957, humanity opened up a new realm in which it expanded its activities.[1] In the same way, after the diffusion of the internet, a new realm defined as 'cyberspace' came to be added in the realms of human activity. As though the technological evolution unlocks new fields of activities, it becomes clear that the new fields require appropriate rules and regulations in order to avoid legal anarchy.

Among the different legal questions to be answered, one of the most pressing is who has jurisdiction in the new domains. Even if both outer space and cyberspace have their own particularities which complicate the question of 'if' and 'how' states can exercise jurisdiction, having a defined legal regime works in the best interest of

---

[1]Nasa, Steve Garber, 'Sputnik and the Dawn of the Space Age' (*NASA History website*, 10 October 2007) <https://history.nasa.gov/sputnik/> last accessed 14 March 2021.

A. E. Farsaris (✉)
University of Luxembourg, Mondercange, Luxembourg

A. Froehlich (ed.), *Outer Space and Cyber Space*, Studies in Space Policy 33,
https://doi.org/10.1007/978-3-030-80023-9_1

all the actors. This chapter will analyse and compare the realms of outer space and cyberspace in light with the applicable jurisdiction regime highlighting the similarities and connections.

Before however doing so, it is appropriate to offer some definitions.

1.2 Definitions

Even if there is no precise definition of where outer space starts beyond the airspace in international law,[2] it is not necessary to provide a definition for the purposes of this article, as the activities concerned, fall well within what is generally considered as 'outer space', such as activities in Earth's orbit or on celestial bodies. On the other hand, it is appropriate to provide a definition of 'cyberspace', as well as of 'state jurisdiction' in order to better understand its applicability to the new realm of human activities.

1.2.1 What is Cyberspace?

Defining cyberspace has raised questions concerning its legal, technical, political and philosophical nature.[3] From a legal perspective however, there is no definition explicitly referring to cyberspace in international law. Nevertheless, in more general terms, cyberspace can be defined as 'a global domain within the information environment whose distinctive and unique character is framed by the use of electronics and the electromagnetic spectrum to create, store, modify, exchange, and exploit information via interdependent and interconnected networks using information-communication technologies'.[4]

It is clear that cyberspace mainly refers to the non-physical space existing between computer nodes.[5] However, based on the definition provided, cyberspace is not only limited in a virtual space completely unrelated to the commonly understood concept of territoriality, but has a rather strong nexus with the physical objects supporting it. Individual devices, servers, satellites, all participate in

---

[2]Bill Warners, 'Patents 254 Miles up: Jurisdictional Issues Onboard the International Space Station' (2020) 19 UIC Rev Intell Prop L 365, 377.

[3]Nicholas Tsagourias, The legal status of cyberspace in Nicholas Tsagourias, Russell Buchan (eds) Research Handbook on International Law and Cyberspace (Edward Elgar Publishing, 2015) 14.

[4]Daniel T Kuehl, From cyberspace to cyberpower: Defining the problem in Franklin D Kramer, Stuart H Starr, Larry K Wentz (eds), Cyberpower and National Security (National Defense University Press 2009) 28.

[5]See The Law.com Dictionary <https://dictionary.thelaw.com/cyberspace/> last accessed 14 March 2021.

bringing the cyberspace together. This is particularly important in order to assess its status under international law in the following sections.

Having offered a basic definition of cyberspace and of what it consists of, the next step would be to provide a definition for the next important term in this analysis.

1.2.2 State Jurisdiction

Jurisdiction is considered to be the 'legal instantiation' of sovereignty.[6] Hence, it is important to refer first to the concept of sovereignty before analysing jurisdiction.

Sovereignty is historically associated with the control of a territory. Its emergence as a legal concept coincided with the apportionment of territories and with the political and legal recognition of such territorial compartmentalisation by the Treaty of Westphalia.[7] Therefore, a part of scholars believe that cyberspace should not be subject to it.[8] For the scholars adopting this view, in addition to the 'borderless' and 'a-territorial' nature of cyberspace, the fact that traditionally sovereignty is essentially a form of power and legitimacy bound to its effects in territorially based entities, should mean that cyberspace should remain outside of it, developing its own distinct system of self-regulation.[9] However, States can have control over activities that do not enter within a strict concept of territory, but are nevertheless under their jurisdiction.[10] Moreover, there have been views expressing that it is important to disentangle the concept of sovereignty from territory. Indeed, territory has been described as 'a container of power' in contrast to just 'a place where powers are located'.[11] In essence therefore, sovereignty is based on power and not on territory, meaning that its application goes beyond any allocated territory and reaches non-territorial entities as well.[12]

Even if the opinion on sovereignty being attached to the concept of power rather than the concept of territory, is not dominant, cyberspace can nevertheless affect physical installations within a State's territory. Thus, cyberspace includes, to some extent, consequences which affect the territory of states and could fall within their jurisdiction.

---

[6]James Crawford, Brownlie's Principles of Public International Law (OUP, 2012) 456.

[7]Tsagourias (n 3) 17.

[8]David R Johnson, David G Post, 'Law and borders: The rise of law in cyberspace' (1996) 48 Stanford L Rev 1367.

[9]Tsagourias (n 3) 16 citing Johnson, Post Ibid.

[10]Prime examples are the jurisdiction provisions in the High Seas and Outer Space where States exercise jurisdiction on a basis of quasi-territoriality; see s 3.

[11]Peter J Taylor, 'The state as container: Territoriality in the modern world-system', (1994) 18 Progress in Human Geography, 151.

[12]Tsagourias (n 3) 18.

Having explained how the essence of sovereignty can apply to cyberspace as well, it is now time to analyse jurisdiction. Jurisdiction in the international context can be defined broadly as the interest of states in obtaining and maintaining jurisdiction over their nationals, territory and acts that affect their broader welfare.[13] Generally, the notion of jurisdiction comprises three kinds of power: first, the power to prescribe (referring to the power of a government to prescribe sanctions), second, the power to adjudicate (referring to the power of the State's courts to hear disputes) and third, the power to enforce (referring to the government's power to compel compliance or punish noncompliance with the laws).[14]

In order to establish jurisdiction, the international community has generally adopted five principles: territorial, active nationality, passive nationality, protective and universal. Starting with the principle of territorial jurisdiction, it is rooted in the notion of sovereignty of a nation state.[15] It can be divided in two types, subjective and objective. Jurisdiction based on subjective territoriality is accorded to a state over all offences committed in its territory.[16] In a similar way, objective territoriality seeks to protect the same interests but permits the finding of jurisdiction over conducts outside the territory of a state but causing an effect therein.[17]

Under the active nationality principle, which is particularly important in civil law countries, the national state can exercise jurisdiction over crimes committed abroad by its nationals. On the other hand, the passive nationality principle grants jurisdiction to national state of the victim of a crime.[18] Although this principle makes sense intuitively, many states, including the US, have rejected this principle as a basis of jurisdiction.[19]

Another basis for criminal jurisdiction is the protective principle. This principle permits a state to punish offences committed wholly outside its territory when those offences threaten the country's security, integrity or sovereignty.[20] It has been criticised though by some to be overreaching since it allows to extent jurisdiction over conducts that merely pose a threat to a nation.[21] The fact however that it is limited only to certain circumstances, helps to reduce the danger of overreaching.[22]

At last, the final basis for criminal jurisdiction is the universality principle. It confers jurisdiction to every state for certain categories of crimes because they are

[13]Stacy J Ratner, 'Establishing the Extraterrestrial: Criminal Jurisdiction and the International Space Station' (1999) 22 B C Int'l & Comp L Rev 323, 328.

[14]Ahmad Kamal, The Law of Cyber-Space, An Invitation to the Table of Negotiations (United Nations Institute of Training and Research, UNITAR, 2005) 197.

[15]A J Young, Law and Policy in the Space Stations' Era (Martinus Nijhoff Publishers 1989) 152.

[16]Covey T Oliver, The International Legal System (1995) 133.

[17]Ratner (n 13) 329.

[18]Hans P Sinha, 'Criminal Jurisdiction on the International Space Station' (2004) 30 J Space L 97.

[19]Ratner (n 13) 329.

[20]Ibid 330.

[21]Ibid.

[22]Ibid.

deemed especially offensive to the international community as a whole, such as piracy and international crimes (war crimes, crimes against humanity, genocide). The prosecution of Adolph Eichmann by the State of Israel in 1962 is generally seen as an exercise of universal criminal jurisdiction.[23]

1.3 An Overview—Jurisdiction in Outer Space

At this stage, before comparing cyberspace to outer space, it is useful to analyse how the international community tackled the issue of jurisdiction in the latter when the issue of human activities in space first appeared.

Even before the first launch of an object into outer space, most scholars and states were concerned about the legal regime applying to this new realm 'unlocked' by the rapid growth of technology.[24] In order to establish the applicable regime, analogies were quickly made comparing outer space with other common places such as the High Seas, Antarctica and the airspace.[25] Whereas the regulation of the airspace where a state retains 'complete and exclusive sovereignty over the airspace over its territory'[26] was soon abandoned as a solution, the regime established in outer space concerning sovereignty and jurisdiction was similar to the one in the High Seas or Antarctica. Indeed, outer space was left outside the sovereignty of states and was described as 'the province of all mankind'.[27] However, the lack of sovereignty does not mean that states are precluded exercising any sovereignty rights.

Indeed, even if the Outer Space Treaty (OST)[28] sets space aside as an extra-jurisdictional territory, this does not limit states from exercising state jurisdiction and control 'over people, institutions, and objects in outer space',[29] under a quasi-territorial regime.

---

[23]Attorney General of the Government of Israel v Eichmann 16 PD 2033 (1962).

[24]See Stephan Hobe, Niklas Hedmasn, Preamble in Hobe, Schmidt-Tedd, Schrogl (eds) Cologne Commentary on Space Law – Outer Space Treaty.

[25]R Jakhu, S Freeland, Article II in S Hobe, B Schmidt-Tedd, K U Schrogl, G M Goh (eds), Cologne Commentary on Space Law: Volume I (Carl Heymanns 2009) 44–63; Frans von der Dunk, International Space Law in Frans vor der Dunk, Fabio Tronchetti (eds), Handbook of Space Law (Edward Elgar Publishing Ltd 2015) 56.

[26]Convention on International Civil Aviation (adopted 7 December 1944, entered into force 4 April 1944) UNTS 295 art 1.

[27]Treaty on Principles Governing the Activities of States in the Exploration and Use of Outer Space, including the Moon and other Celestial Bodies, 19 December 1966 (entered into force 10 October 1967) 610 UNTS 205 (OST) art I.

[28]Ibid.

[29]Stephen Gorove, Legal Problems of Manned Spaceflight in Chia-Jui Cheng (ed), The Use of Airspace and Outer Space for all Mankind in the 21st Century (Kluwer Law International 1995) 246.

The OST itself provides for the first binding principles regarding jurisdiction. It recognised that states have jurisdiction over objects launched into space. Indeed, Article VIII provides for 'jurisdiction and control' of the state on whose registry a space object is launched into outer space, over such object and its personnel. The state shall retain jurisdiction whilst the space object is in outer space or on celestial bodies. The same Article further specifies that a state shall also retain ownership over space objects regardless of their presence in outer space, on celestial bodies or of their return to the Earth. The provisions laid out in the Treaty, hence, grant the authority to states to exert jurisdiction over objects and individuals, limited only to the registered space objects and their personnel. States retain a quasi-territorial jurisdiction, similar to that for ships and aircraft.[30]

The OST was followed by the Registration and Liability Conventions, which also contained provisions concerning outer space jurisdiction, confirming the quasi-territorial regime. The Registration Convention requires states to register space objects that they launch into outer space 'in an appropriate registry'.[31] It is the state of registry that exercises jurisdiction upon those objects. It also adds to the regime provided by the Outer Space Treaty that when there are multiple launching states, such states may allocate jurisdiction over the space objects through agreements between them.[32] The Liability Convention repeats the same principle of jurisdiction. Article VIII of the Liability Convention confirms the state of registry's jurisdiction over all of its space objects.

The provided space law transforms space objects into pieces of quasi-territory of a particular state, inviting the concurrent exercise of jurisdiction.[33] The regime follows the rules of jurisdiction '*ratione instrumenti*' to cover space objects in the same manner the principle was already extended to aircraft and ships.[34] Indeed, as it was already established in the Law of the Sea, the state of registry (flag state) of an object under quasi-territorial regime, retains jurisdiction regardless of where the object is found.[35] Additionally, in the same manner that a state cannot claim an aircraft that has landed on its territory, it cannot claim a registered space object that happened to return from space to its territory.[36]

---

[30]See United Nations Convention on the Law of the Seas (done 10 December 1982, entered into force 16 November 1994) 1833 UNTS 397 (UNCLOS) arts 86–120; P G Dembling, D M Arons, 'The Evolution of the Outer Space Treaty' (1967) 33 Journal of Air Law and Commerce 429–432.

[31]Convention on Registration of Objects Launched into Outer Space (adopted 12 November 1974, entered into force 15 September 1976) UNGA Res 3235 (XXIX) art II.

[32]Ibid.

[33]W Zhang, 'Extraterritorial Jurisdiction on Celestial Bodies' (2019) 47 Space Policy 148, 150.

[34]Gbenga Oduntan, Sovereignty and Jurisdiction in the Airspace and Outer Space, Legal Criteria for Spatial Delimitation, (Routledge 2012) 180.

[35]UNCLOS (n 30) art 94 'Every state shall effectively exercise its jurisdiction and control in administrative, technical and social matters over ships flying its flag'.

[36]Oduntan (n 34).

Another legal document providing for outer space jurisdiction, albeit not with significant binding value, is the Space Protocol[37] on the Cape Town Convention.[38] The Cape Town Convention is a private international law treaty intended to standardise transactions involving mobile equipment of high-value or particular economic significance. The treaty with its Space Protocol provides for a set of rules concerning the prioritisation, protection and enforcement of rights and interests in movable space property.[39]

The Space Protocol allows for different states to exercise jurisdiction upon a space asset. Its Article XXII details different possibilities of jurisdiction over the same space asset:

> The courts of a Contracting State: (i) in which the space asset is situated; (ii) from which the space asset may be controlled; (iii) in which the debtor is located; (iv) in which the space asset is registered; (v) which has granted a license in respect of the space asset; or (vi) otherwise having a close connection with the space asset.

Such states, 'in accordance with the law of the contracting state' shall 'co-operate to the maximum extent possible' to assist a creditor exercise his remedies under the Cape Town Convention and the Space Protocol.[40]

In contrast with the Outer Space Treaty, the Space Protocol gives a broad list of states that could exercise jurisdiction upon a space asset. As a matter of fact, any state 'having a close connection' with the space asset can exercise jurisdiction. The inconsistencies between the two legal instruments could be explained by interpreting Article VIII of the Outer Space Treaty as a nonexclusive provision.[41] Indeed, following this interpretation, a signing state of registry to the Space Protocol, can decide based on international law to transfer some of its jurisdiction granted by Article VIII. The Cape Town Convention with its Space Protocol does not breach the provisions of the Outer Space Treaty, it rather supplements the outer space jurisdiction regime.

The Space Protocol, unlike the Cape Town Convention itself, has not been ratified by the parties of the Convention and has not entered into force yet.[42] Even though the Protocol is not of binding importance, it is clear that the limited jurisdiction provisions set forth by the UN space treaties, do not create a sufficient jurisdictional regime for the plethora of space activities. The rules of jurisdiction in

---

[37]Protocol to the Convention on International interests in Mobile Equipment on Matters Specific to Space Assets (signed 9 March 2012) (Space Protocol).

[38]Cape Town Convention on International interests in Mobile Equipment (signed 16 November 2001, entered into force 1 March 2006) 2307 UNTS 285.

[39]M Sundahl, Cape Town Convention: Its Application to Space Assets and Relation to the Law of Outer Space (Martinus Nijhoff Publishers 2013) 171; Zhang (n 86) 151.

[40]Zhang (n 33) 151.

[41]See B Cheng, 'The Extra-Terrestrial Application of International Law' (1965) 18 1 Current Legal Problems 132–152.

[42]As of 2020 no state has ratified the Space Protocol <https://en.wikipedia.org/wiki/Cape_Town_Treaty> last accessed 14 March 2021.

outer space are not exclusive, subsequent agreements can integrate the jurisdictional bases to exercise legal control in space.

Another important document that provides a jurisdiction regime in outer space is the Intergovernmental Agreement (IGA) for the International Space Station (ISS).[43] Article 5 provides the general jurisdiction principle regarding aboard the ISS. First, Article 5 entails that each of the partner states retains jurisdiction over the 'flight elements' they register. Hence, it opts for a 'quasi-territorial' regime. Second, Article 5 also allows a state party to exercise jurisdiction over personnel 'in or on the Space Station' who are its nationals. This provision follows the nationality principle where a state can exercise jurisdiction over any of its nationals abroad. However, the most interesting provision in the IGA is Article 22.

Article 22 dealt with criminal jurisdiction for the first time in space law. Its first paragraph states that the partners 'may exercise criminal jurisdiction over personnel in or on any flight element who are their respective nationals'. Pursuant to this provision, the partner states retain jurisdiction over their nationals regardless of on which flight element a national commits an offence, following, hence, the nationality principle of jurisdiction. Paragraph 2 of Article 22 broadens the possibilities of exerting jurisdiction beyond the nationality principle, and hence, it allows to exercise jurisdiction over non-nationals under certain circumstances. It provides that a partner state can exercise criminal jurisdiction over an offence committed on board the ISS, when such offence 'affects the life or safety' of its nationals or 'occurs in or on or causes damage' to its flight elements regardless of the perpetrator's nationality. It seems therefore that the partners chose both the territoriality and the passive nationality principles as a basis for jurisdiction as enshrined in paragraph 2.

Even though the drafters must have viewed the possibility of criminal acts occurring on the ISS as a remote one given that the crew on board is composed by highly trained military or scientific personnel, they were concerned about establishing a regime for criminal jurisdiction. This is a case where law provides regulations for certain circumstances before such circumstances even arise. As Haley had stated in the early development days of space law, 'law must precede man into space'.[44]

This however has not been the case concerning cyberspace as there is still no international legal regime regulating the activities, even if it has been active for decades. At this stage, cyberspace should be analysed in comparison with outer space regarding the similarities in a possible international jurisdiction regime.

---

[43]Agreement Among the Government of Canada, Governments of Member States of the European Space Agency, the Government of Japan, the Government of the Russian Federation, and the Government of the United States of America concerning Cooperation on the Civil International Space Station, 29 January 1998 (entered into force 27 March 2001) (IGA).

[44]Andrew G Haley, Space Age Presents Immediate Legal Problems in Proceedings of the First Colloquium on the Law of Outer Space [1959].

1.4 Cyberspace and Outer Space—A Jurisdictional Parallel?

When one thinks of cyberspace in comparison to outer space, there are definitely fundamental differences. First and foremost, the very nature of cyberspace is not only different to outer space, but to any other common place that has been regulated in international law. Indeed, cyberspace is a human-made borderless environment purely under human control, whereas outer space is a natural one. Other major differences include the 'threshold of entry' needed, since almost everyone with an internet connection has access, and the fact that in outer space, the activities of privates are almost non-existent when comparing to cyberspace. The diverse nature of cyberspace from any other regulated area poses the question as to what is its nature under international law.

1.4.1 Cyberspace as Global Commons

In spite of the differences of cyberspace with outer space, the High Seas and Antarctica, there have been views including it between the global commons. Analysts have suggested that the defining character of a commons is the question of freedom of access.[45] According to this, the free access nature of cyberspace could include it within the realms of global commons. This has been expressed by the NATO document 'Assured Access to the Global Commons', which asserts the importance of the assured access and use of 'the maritime, air, space, and cyber-space domains that are the commons'.[46]

On the other hand, however, cyberspace cannot be designated as a global commons without the consent of States and the agreement on the rules and principles that govern it. This is another major difference between cyberspace and outer space, as the OST had defined the rules applying to space and created an authoritative reference point soon after the first space activities. On the contrary, there has been no similar treaty to define cyberspace and enact the cooperation instruments between the States.[47] It is for these reasons that cyberspace has been described as an 'imperfect commons'.[48]

---

[45]Paul Meyer, Outer Space and Cyberspace: A Tale of Two Security Realms in Anna-Maria Osula, Henry Rõigas (eds) International Cyber Norms: Legal, Policy & Industry Perspectives (NATO CCD COE Publications, 2016) 157.

[46]Mark Barrett, et al., Assured Access to the Global Commons (Norfolk: North Atlantic Treaty Organization, 2011), xii, <www.act.nato.int/images/stories/events/2010/gc/aagc_finalreport.pdf> last accessed 14 March 2021.

[47]Meyer (n 45) 160.

[48]Tsagourias (n 3) nn 92 citing Joseph S Nye, The Future of Power (PublicAffairs 2011) 143.

1.4.2 Similarities with Outer Space

When defining a new concept, international law is influenced by the understanding of how spaces have been previously regulated. International lawyers indeed, will look for analogies with other physical spaces and the law principles applying to them.[49] When considering cyberspace and its similarities to physical spaces, outer space offers a useful paradigm.

Indeed, in spite of the obvious differences, it is interesting to analyse the similarities for what jurisdiction is concerned. Outer space consists of the 'open space', as well as its 'territorial' elements such as the Moon and the rest of the celestial bodies.[50] However, for what concerns jurisdiction purposes, since no state can exercise sovereignty rights and jurisdiction or make any territoriality claims on the Moon and the celestial bodies, it could be considered a-territorial to a certain degree.[51] Obviously there is still the 'open space' around Earth's orbit and in between the celestial bodies, but the jurisdictional provisions require human-made vehicles and installations in order for such rules to apply.

This is in a big degree similar to the nature of cyberspace and the applicable jurisdiction rules. As it was referenced earlier, a jurisdictional regime applying to cyberspace could enact rules based on the territorial elements and effects of cyberspace, very similar to the jurisdictional provisions in outer space which affect only the quasi-territorial human-made elements.[52] For instance, a provision concerning a criminal offence in cyberspace could grant jurisdiction over it based on a territorial basis of where the offence was committed.

Having demonstrated the similarity on the 'territorial connection' of outer space and cyberspace, the jurisdiction principles applying in outer space should be appropriate, to a certain degree, to establish jurisdiction in cyberspace as well. As referenced in section 2, jurisdiction in international law is based on the classic jurisdictional principles. Starting with the territoriality principle, states retain jurisdiction over the space objects they registered on a quasi-territorial basis. On the same basis, the territoriality principle could apply with regards to cyberspace for what concerns the 'physical' element of internet operations, or the location of the offender in cases of cybercrime. Given though the borderless nature of cyberspace and the difficulty to assess certain times where the damage was caused from, the territoriality principle would not be sufficient to provide the jurisdictional basis by itself.

As the ISS partners came to realise in the drafting of the 1998 IGA, the territoriality principle was not ideal for ISS based jurisdiction given the borderless nature of the station, where different modules registered by different partner states are set together to provide a common environment for the personnel.[53] A set of

[49]Ibid 15–16.

[50]OST (n 27) art I.

[51]Ibid art II.

[52]Ibid art VIII; IGA (n 43) art 5, 22.

[53]Ratner (n 13) 335.

principles better suited to the environment of outer space, as well as for cyberspace are the nationality and passive personality principles. Indeed, given the borderless nature of cyberspace, establishing a jurisdiction regime based on the nationality of the offender or even the nationality of the victim offers a much more simplified approach.

Nevertheless, the legal regime of cyberspace is not a legal vacuum as there have been steps towards establishing an international cooperation in its regulation.

1.5 Cyberspace—Existing Legal Initiatives

The most meaningful legal tool for international cyberspace regulation is the 2001 Budapest Convention on Cybercrime.[54] The Budapest Convention was drafted by the Council of Europe and was opened for signature for states beyond its member states, aiming to establish a common criminal policy in order to protect the society against cybercrime.[55] Among the different provisions, Article 22 was dedicated to the question of jurisdiction.

Paragraph 1 of Article 22 establishes the jurisdiction regime and grants states the right to adopt measures to establish jurisdiction for offences committed on their territory (including ships and aircraft), and for offences punished by criminal law when committed by their respective nationals outside their territory. It is obvious that Article 22 implements the territoriality and personality principles of jurisdiction over cybercrimes. Moreover, for the cases when more than one states claim jurisdiction over an alleged offence, paragraph 5 establishes the obligation for such states to consult 'where appropriate' in order to determine the most appropriate jurisdiction for prosecution.

Whilst limited only to provisions aimed to fight against cybercrime and not achieving vast acceptance when compared to the OST, the Budapest Convention remains the most significant international legal document in cyberspace.[56]

Another interesting document to note is the invitation to the table of negotiations published by the United Nations Institute of Training and Research entitled 'The Law of Cyber-Space'.[57] This document aims to invite the member states of the United Nations to consider the importance of starting in a sector that had been ignored in the previous years.[58] Unlike the Budapest Convention, it concentrates its analysis not only on the concept of cybercrime, but on a vast array of issues

---

[54]Council of Europe Convention on Cybercrime ETS No 185 (entry into force 01 July 2004) (Budapest Convention).

[55]Ibid Preamble.

[56]As of December 2020, there are 65 parties to the Budapest Convention, see <https://en.wikipedia.org/wiki/Convention_on_Cybercrime> last accessed 14 March 2021.

[57]Kamal (n 14).

[58]Ibid preface.

including the right to access, data protection, intellectual property and, for what concerns this paper, jurisdiction.

Regarding jurisdiction, it recognises the 'loopholes' of the current regime and the need to establish further cooperation. However, it also recognises that the issues that mostly affect jurisdiction in cyberspace are criminal offences. Therefore, as explained above, the international regime already offers a legal instrument regarding cybercrime. Thus, the need would not be to establish a complete regime concerning criminal jurisdiction in cyberspace, but rather achieve greater cooperation in the issues that still remain vague and problematic.[59]

The latter developments also show that the trend shifts towards an international regulation of cyberspace. Following the 2018 General Assembly Resolution which aimed to collect the countries' views on cybercrime, discussions arose concerning a cybercrime treaty, which would essentially replace the Budapest Convention which has received criticism as being too sovereignty-intrusive.

1.6 Some Considerations

In spite of the similarities with outer space for what regards the jurisdictional basis, cyberspace has different legal needs. Indeed, even if outer space is becoming increasingly accessible and the private sector begins to take the helm from states for what concerns the number of activities, its whole legal body remains state-based. On the other hand, cyberspace and the internet were from the very beginning open to everyone, reaching every country and counting billions of users. The state of 'free access' in cyberspace and its borderless nature make it difficult to trace crimes, which could result in a realm that in the absence of international legal coordination, could become the centre for illegal activities.

Again, on the contrary with outer space, the main legal issue concerning jurisdiction in cyberspace are criminal offences. The drafters of the OST or of the following international space law treaties were not concerned of the possibility of criminal offences being committed in a realm where only highly trained government officials had access. Even if with the development of the technology such concerns begun to arise starting with Article 22 of the ISS IGA, it is not yet comparable to cyberspace. Therefore, an international legal regime in cyberspace would be sufficient for jurisdictional purposes if it would cover criminal jurisdiction and all its implications. The Budapest Convention is a positive example to this effort, however, it has not received the necessary acceptance. Indeed, with only 65 parties to the Convention, several state actors remain outside including Russia, China or India.[60]

[59]Ibid 204.

[60]See parties to the Convention (n 56).

At this point, given the importance of international cooperation and coordination against cybercrime, a UN treaty that could achieve a high acceptance and global representation appears to be the solution.

1.7 Conclusion

Access to outer space was only achieved after the development of the new technologies. As it arose, states agreed on the necessity to establish a legal regime to achieve its exploration, but at the same time, protect the common good. On the other hand, cyberspace was also accessed by the development of the technology, but its regulation did not follow the same route.

In spite of spatial and legal similarities between the two realms, cyberspace regulation faces different needs than any other previously regulating experience in international law. It is for this reason its legal regime cannot be established in parallel with the regime of outer space, nor be developed only on its precedents.

Whereas an analogy with outer space is helpful in building an understanding that, in its turn, aids in the choice of the base rules, a realm as particular as cyberspace requires the creation of a new legal paradigm.

Alexandros Eleftherios Farsaris is currently on his thesis for the Master on Space, Communication and Media Law at the University of Luxembourg. In the past years he has conducted research on space law and a thesis on the topic. He is aspired to continue the research in the field of international space law.

Chapter 2
Extraterrestrial Netizenship: Citizenship in Outer Space and Cyberspace

Connor Hogan

> *People sort of think the concept of a human community off this Earth is hundreds and hundreds of years away. And it's really not. We really have the opportunity to do this right, and to promote peace through individual behaviour.*
>
> —Prof. Michelle Slawecki Hanlon, Co-Founder and the President of For All Moonkind, Inc.

Abstract Human beings in outer space currently hold the citizenship of their origin country, as their individual presence beyond Earth is transient, highly specialised and purpose driven. However, as more people begin to live and work in outer space on a permanent basis, questions of citizenship are due to gain more salience, particularly as space remains a 'global commons' outside of the international state system. In this regard, there are important comparisons to be drawn with the legal environment of cyberspace, in which similar challenges are already being faced. This chapter will investigate the similarities between the two realms and propose that the international community should make preparations for new forms of cosmopolitan citizenship, commensurate with the unique environments of outer space and cyberspace.

C. Hogan (✉)
School of Politics and International Relations (SPIRe), University College Dublin (UCD), Dublin, Ireland
e-mail: connor.hogan@ucdconnect.ie

A. Froehlich (ed.), *Outer Space and Cyber Space*, Studies in Space Policy 33,
https://doi.org/10.1007/978-3-030-80023-9_2

2.1 Introduction

The 21st century is set to see an increase in the amount of people living and working in space on a permanent basis. Several developments give us reason to suppose this: the ESA's recruiting of astronauts for the first time in a decade,[1] NASA's continued development of its Artemis lunar program, China's plans to build a new space station and send taikonauts to the Moon by the end of the decade, the rise of private space actors such as *SpaceX*, and legislation in the United States and Europe in anticipation of asteroidal mining in the medium term.[2] Currently, visitors to space are virtually all scientists conducting research on a temporary basis (6–12 months) and are legally under the jurisdiction of their home nation. As this population is marginal, extremely specialised, and transient, the existing body of international space law has been able to accommodate their status without contention; the 'extraterritoriality' of an astronaut's citizenship status is accepted. Article II of the Outer Space Treaty of 1967 stipulates that outer space is not subject to national appropriation by claim or by occupation, and thus in practice, visitors bring the citizenship of their state of origin to this region beyond the international state system, on the assumption that they will shortly return to their sovereign state. However, if the aforementioned developments continue apace, it is conceivable that entire communities of people will be living and working in space on a much more permanent basis within the next few decades. With such communities will inevitably come inter alia complex legal questions of individual jurisdiction, work permits and migration, and thus citizenship. Thus, it is crucial for scholars of space policy, law, and political theory to begin work on a conceptual framework for citizenship in outer space, and it is in this context that the example of cyberspace can offer valuable insights. Like outer space, cyberspace exists 'beyond the claims of sovereign states and national appropriation',[3] and is adapting to the unique legal and political challenges such status entails. However, whereas the population of outer space is for now relatively low, there are an estimated 3.2 billion 'netizens',[4] and thus the exact relationship between individuals, their state of origin, the international community and cyberspace is already being debated. This chapter will survey the current legal framework for citizenship in both these cases, investigate important similarities and challenges, and use a typology of citizenships

[1]ESA, 'N°3–2021: Call for media: ESA seeks new astronauts' (*European Space Agency (ESA) website*, 8 February 2021) <https://www.esa.int/Newsroom/Press_Releases/Call_for_media_ESA_seeks_new_astronauts_-_applications_open_31_March_2021> accessed 10 March 2021.

[2]I Christensen and others, 'New policies needed to advance space mining' [2019] 35(2) Issues in Science and Technology 26–30.

[3]Paul Meyer, Outer Space and Cyber Space A Tale of Two Security Realms. in Anna-Maria Osula and Henry Rõigas (eds), International Cyber Norms: Legal, Policy & Industry Perspectives (NATO CCD COE Publications 2016) 155.

[4]Hao Yeli, 'A Three-Perspective Theory of Cyber Sovereignty' [2017] 7(2) PRISM 111.

(extraterritorial, supplementary, cosmopolitan) to propose a new model for both outer space and cyberspace: cosmopolitan netizenship, and extraterrestrial cosmopolitan citizenship.

2.2 Contemporary and Future Citizenship

Citizenship in its contemporary form is primarily a Western phenomenon that originated in Europe around the Middle Ages (albeit with roots stretching back as far as Rome and Ancient Greece if not earlier).[5] Although throughout its modern history it was often applied unevenly (or violently imposed), citizenship has for all intents and purposes become a ubiquitous phenomenon of human life on Earth: with some exceptions, almost everyone is a citizen of somewhere. For the purposes of this chapter, I will define citizenship as 'status of membership in a particular political community that entails equal basic rights, legal obligations and opportunities to participate actively in political decision making'.[6] Of course, citizenship itself can (and often does) entail a more maximal array of rights and duties, however the above definition captures the most basal and widely applied definition. Crucially: it highlights that contemporary citizenship is generated and sustained by a sovereign state, in the context of an international state system that recognises it. As human presence expands into regions beyond the world of sovereign states, then, we must ask what should become of citizenship.

To highlight some challenges to the traditional state-citizenship model, let us examine the components outlined in our definition one by one:

2.2.1 Basic Rights

The most standard rights which citizenship entails is the right to live and work in a given country (citizens of the European Union for example have the right to free movement, settlement, and employment across the Union).[7] For outer space, this aspect is for the most part consistent with contemporary extraterritorial (or 'carried') citizenship in the short to medium term, as an individual working in outer space may simply retain the right to also do so in their country of origin. Issues may arise in the longer term, however, as citizens with less and less exposure to their state of origin claim the right to live and work there. In the case of cyberspace, challenges are already arising in the EU and elsewhere concerning the competent

[5]Derek Heater, A brief history of citizenship (NYU Press 2004).

[6]Rainer Bauböck, 'The rights and duties of external citizenship' [2009] 13(5) Citizenship studies 475-499.

[7]EC Treaty (Treaty of Rome, as amended).

courts to deal with employment issues for virtual/remote workers (an increasingly large constituency in the age of COVID-19) whose main employers are abroad and whose work takes place in cyberspace.[8]

2.2.2 *Legal Obligations*

At a minimum, citizenship usually entails an obligation to pay taxes and obey the law.[9] This aspect of citizenship is particularly problematic in the context of a global commons, as it implies some level of territoriality. To whom would future, permanent citizens of space settlements legitimately pay taxes? Similarly: whose laws should they obey? While the current body of international space law offers preliminary answers (see below), a more detailed legal regime will be required as the population of space increases. As with basic rights, there are useful analogues in modern cyber law: jurisdiction can vary according to an individual nations' cyber laws, and thus international harmonisation is an oft stated goal for the field, given the extra-territorial and transnational nature of cybercrime.[10]

2.2.3 *Political Decision Making*

As communities in outer space grow, a political and legal framework for autonomous and semi-autonomous decision making will be increasingly needed, particularly for long missions and remote outposts. Institutions could be created in which to manage these processes in conjunction with the space settlements and their home (or patron) nations, but without new forms of citizenship to connect new supranational institutions to the individuals living within them, their democratic legitimacy is endangered.[11] Therefore, new forms of citizenship which account for the extraterritorial nature of both realms are needed to fulfil and strengthen the political decision making requirement.

---

[8]MT Carinci and A Henke, 'Employment relations via the web with international elements: Issues and proposals as to the applicable law and determination of jurisdiction in light of EU rules and principles' [2020] 20(10) European Labour Law Journal 1–22.

[9]Heater (n 7).

[10]AAS Al hait, 'Jurisdiction in Cybercrimes: A Comparative Study' [2014] 2(1) Journal of Law, Policy and Globalization 75.

[11]A Schlenker and J Blatter, 'Conceptualizing and evaluating (new) forms of citizenship between nationalism and cosmopolitanism' [2014] 21(6) Democratization 1091.

2.3 A Typology of Citizenships

Additionally, there are several modes of citizenship which either currently exist in the context of terrae nullius, or which have been theorised as potential future configurations.[12] With the exception of cosmopolitan citizenship, these are not "new" citizenships per se, but rather special applications of national citizenship for situations beyond the international system of sovereign states—and thus of great interest as we formulate new forms for outer and cyberspace. I will briefly survey these before outlining the current framework for individuals in outer space and cyberspace.

2.3.1 Extraterritorial

This is the form of citizenship that currently pervades not only in outer space but also in Antarctica, and to ambassadors and some diplomatic officials. An individual in these situations is a *representative* of their origin state and carries their national citizenship with them. The rights, duties and obligations are 'negotiated between states and non-resident citizens'.[13] Persons born in this context generally (though not always) assume the citizenship of their parents. In this context, it is assumed that most legal disputes will fall under the jurisdiction of a given individual's home nation (in conjunction with international treaties), however as illustrated above, this will become increasingly problematic in multi-national, permanent settlements with a complex, multi-person nexus of rights, duties and obligations that discrete national legal regimes will struggle to absorb and arbitrate.[14] A useful case study from cyberspace in this regard is the 'digital diasporas'[15] of certain developing countries, where non-resident citizens have created a virtual, extra-territorial public sphere outside of the origin nation, with contested (or potentially renegotiated) rights and duties.

---

[12]There are additional forms (such as dual citizenship) which I will not discuss here, as they describe specific forms on interstate citizenships that are not applicable to outer space or cyberspace.

[13]M Collyer, 'A geography of extra-territorial citizenship: Explanations of external voting' [2014] 2(1) Migration Studies 56.

[14]WL Zhang, 'Extraterritorial jurisdiction on celestial bodies' [2019] 47 Space Policy 148–157.

[15]A Everett, Digital diaspora: A race for cyberspace (SUNY Press 2009) 10–8.

2.3.2 Supplementary

EU citizens currently hold a *supplementary* citizenship that acts in tandem with their national status and affords them special rights in all EU member states.[16] The advantages of this approach to outer space settlements would be (as in the EU) legal harmonisation of basic rights and obligations, however it is dubious whether this could be practically done without some form of sovereign political institution behind it, in violation of the Outer Space Treaty. In the study of cyberspace, the term 'digital citizen' has been coined to describe those who -regardless of nationality- use the internet in accordance with broadly defined political duties and responsibilities.[17] Building on this, we could consider a form of 'supplementary digital citizenship', in addition to a person's national citizenship, with negotiated competences and precedence between the two (as with EU citizenship).

2.3.3 Cosmopolitan Citizenship

One of the most intensely theorised alternative models of citizenship comes from the school of cosmopolitanism, some of the earlier expressions of which can be found in the writings of Kant and his conception of a 'world citizenship'.[18] Building on formative discussions of legal cosmopolitanism, a Cosmopolitan Approaches to International Law (or CAIL) as an alternative to the Third World Approaches to International Law (or TWAIL),[19] has been proposed in order to assess space governance issues—particularly the inclusion of the legal right to benefit from space and the inclusion of developing actors. Cosmopolitan citizenship is premised on the idea of individuals partaking in political decisions at multiple levels above the national, up to the planetary level,[20] and much of the new cosmopolitan literature suggests a form of 'planetary consciousness'[21] that could serve as the basis for a new model of citizenship for the global commons.

[16]With a relative lack of duties – the 'de-dutification' in EU membership is part of a global trend across the democratic world. See: D Kochenov, 'EU citizenship without duties' [2014] 20 (4) European law journal 482-498.

[17]MK Heath, 'What kind of (digital) citizen?' [2018] 35(5) The International Journal of Information and Learning Technology.

[18]Andrew Linklater, 'Cosmopolitan citizenship' [1998] 2(1) Citizenship studies 23–6.

[19]Timiebi Aganaba-Jeanty, 'Introducing the Cosmopolitan Approaches to International Law (CAIL) lens to analyze governance issues as they affect emerging and aspirant space actors' [2016] 37(1) Space Policy 3–11.

[20]Derek Heater, Citizenship: The civic ideal in world history, politics and education (Manchester University Press 2004).

[21]T Jazeel, 'Spatializing difference beyond cosmopolitanism: Rethinking planetary futures' [2011] 28(5) Theory, Culture & Society 75–97.

Having now outlined the main challenges confronting contemporary citizenship in a global commons beyond the realm of traditional state sovereignty, and having introduced a typology of theorised alternatives, I will now survey the current legal framework regarding citizenship, outer space and cyberspace.

2.4 Current Legal Framework

2.4.1 Space and Celestial Bodies

Currently, the international treaties which govern outer space do not make reference to citizenship, although much of the contemporary discussion in the field concerns state sovereignty. Shortly after the launch of Sputnik-1 in 1957, an international norm was established for outer space which emphasised scientific use and limitation of arms, which has formed the bedrock of the field ever since. Article II of the Outer Space Treaty (OST) of 1967, states that:

> Outer space, including the moon and other celestial bodies, is not subject to national appropriation by claim of sovereignty, by means of use or occupation, or by any other means.[22]

Other widely ratified space treaties (including The Rescue Agreement of 1968, The Space Liability Convention of 1972 and The Registration Convention of 1976) similarly uphold the extraterritorial nature of space and make no reference to citizenship. In practice, this has meant that bodies like the Moon remain a terra nullius (the 'common heritage of mankind'—CHM)[23] in international law, and extraterritorial citizenship for human beings in these environments is assumed: citizens carry their national citizenship to space and back, and are in essence bound to the international treaties in the same way as their home state is.[24] Potential challenges to the CHM (such as former US President Trump's executive order concerning mining on the Moon and asteroids)[25] indicate that further clarity on these principles is needed. Moreover, whereas the relatively small number of actors in the early days of space law meant that more ambiguous or complex disputes could be solved on an ad hoc basis, the multiplicity of both state and private actors now entering the industry (and the increasing complexity of legal processes:

---

[22]United Nations Office for Outer Space Affairs, *International Space Law: United Nations Instruments* (United Nations 2017) 4.

[23]V Pop, Who Owns the Moon? (Springer, Dordrecht 2009) 2.

[24]R Singh and others, Current developments in air space law (National Law University Delhi 2012) 33.

[25]Williams Matt, 'Trump signs an executive order allowing mining the moon and asteroids' (*Universe Today*, 11 April 2020) <https://www.universetoday.com/145622/trump-signs-an-executive-order-allowing-mining-the-moon-and-asteroids/> accessed 12 March 2021.

immigration, births, marriages) will require a more robust judicial framework.[26] A key component of this framework is citizenship.

2.4.2 *Jurisdiction Within Artificial Space Environments*

While outer space and celestial bodies remain the 'common heritage of mankind', objects launched into space (including on-board personnel) are under the jurisdiction of their state of registry. Article VIII of the Outer Space Treaty states that:

> A State Party to the Treaty on whose registry an object launched into outer space is carried shall retain jurisdiction and control over such object, and over any personnel thereof, while in outer space or on a celestial body.[27]

As the treaty makes no reference to the temporality of individual personnel, we can assume that it includes even *permanent* residents of outer space. As stated, although this is a comprehensible and actionable legal principle when the human population of outer space is small, primarily scientific and transient, the status of people living and working permanently on board future settlements will need to be considered and is not currently specified in any current legislation. However, given the multitude and complexity of legal issues this will entail, a more in-depth framework should be formulated, a key aspect of which is citizenship.

2.4.2.1 International Space Station Legal Framework

The International Space Station (ISS) is an indispensable contemporary analogue as we anticipate the necessary legal frameworks for future (larger, more permanent and/or more remote) space stations and settlements. The station has been permanently inhabited since November 2000, with its highest population at thirteen, although it usually accommodates between three and six. The station is governed by the Intergovernmental Agreement (IGA) on Space Station Cooperation (1998), an international treaty which stipulates the jurisdiction, obligations, and responsibilities of participant nations. It is also governed by lower-level bilateral Memoranda of Understanding (MOU), which describes these roles in more detail. Article V of the IGA[28] stipulates that:

> Each partner shall retain jurisdiction and control over the elements it registers and over personnel in or on the Space Station who are its nationals.

---

[26]KN Metcalf, 'A Legal View on Outer Space and Cyberspace: Similarities and Differences' [2018] 10(1) Tallinn Paper 4–9.

[27]United Nations Office for Outer Space Affairs (n 23) 5–6.

[28]ESA, 'Intergovernmental Agreement (IGA) on Space Station Cooperation' (*European Space Agency (ESA) website*) <https://web.archive.org/web/20090610083738/http://www.spaceflight.esa.int/users/index.cfm?act=default.page&level=11&page=1980> accessed 15 March 2021.

In essence, this means that personnel on board the ISS are extraterritorial citizens of their home nation and are legally responsible to it. Although European states are treated as one, any EU state may extend their own national laws to personnel aboard the ISS—thus extraterritorial national citizenship precedes the supranational (or supplementary) in this case. Any conflicts *between* the ISS nations are to be resolved through 'application of other rules and procedures already developed nationally and internationally'.[29] As with outer space and celestial bodies, there is an element of ad hoc diplomacy to inter-citizen legal disputes on the ISS.

2.4.3 International Cyber Law

Similarly to outer space, the internet (or cyberspace) is also generally seen as a global commons,[30] in which state sovereignty is 'sundered',[31] and thus the rights, obligations and political power of individuals vis-à-vis their nations of origin are contested. Indeed, space law has been described as the 'older sibling'[32] of contemporary cyber law, as both were in their inceptions representative of 'a new future, in which national borders and terrestrial disputes played no role',[33] but have encountered comparative difficulties in abandoning traditional state-oriented ways of thinking and organising. However, unlike outer space, the multiplicity of active participants (both state and private) and the globalised, diffuse nature of cyberspace (as well as the anonymity inherent to individual users) has meant that effective international legal instruments have been comparatively difficult to agree on, and in practice its status as a global commons is very much contested in international law.[34] The primary debate in the cyber governance literature concerns state sovereignty, with some arguing that it should be extended into the new virtual realm,[35] and others contending that it should remain a global commons.[36]

[29]ESA, 'International Space Station legal framework' (*European Space Agency (ESA) website*) <https://www.esa.int/Science_Exploration/Human_and_Robotic_Exploration/International_Space_Station/International_Space_Station_legal_framework> accessed 12 March 2021.

[30]ML Mueller, 'Against Sovereignty in cyberspace' [2020] 22(4) International Studies Review 780-1.

[31]ibid.

[32]Metcalf (n 28) 2.

[33]ibid.

[34]Andrew Liaropoulos, Cyberspace Governance and State Sovereignty. in GC Bitros and NC Kyriazis (eds), Democracy and an Open-Economy World Order, ed George Bitros & Nicholas Kyriazis Cham: Springer (Springer 2017).

[35]Posch Reinhard, 'Digital Sovereignty and IT-Security for a Prosperous Society' [2006] Informatics in the Future.

[36]Stephane Couture and Sophie Toupin, 'What Does the Notion of "Sovereignty" Mean When Referring to the Digital?' [2019] 21(10) New Media & Society 2305-2322.

There is still no universally agreed international legal principle regarding dispute settlement in cyberspace. The basic rights and legal obligations of individual citizens in cyberspace generally fall under their national courts and are primarily concerned with commerce and crime.[37] The 'citizenship' of an individual engaging in these activities in a virtual space is therefore in many cases extraterritorial: they are a representative of their registry nation, and responsible to its laws. An exceptional case that one might consider is the extraterritorial scope of the GDPR: it applies not only to citizens and businesses within the Union, but can also apply to businesses that have no formal establishment within the EU.[38]

Generally speaking however, cyber law and security remains within the purview of the nation state. Although understandable in the era of international cybercrime, this tenaciously state-oriented approach contradicts the 'spirit of the internet'[39] as a realm of unfettered interconnectivity. In order to assert one's own sovereignty, one needs an exclusive (or near-exclusive) domain over a given realm, in which to impose one's own rules and norms; this is sovereignty by definition. Approaches to cyberspace governance that are based primarily on state sovereignty run the risk of *balkanising* the internet, by necessitating discrete *national* cyberspaces over which individual states may hold sway. Given the global and open nature of the internet up until this point, such speculation may seem far-fetched, however consider that contemporary Chinese 'netizens'[40] already do not for the most part exist in the same cyberspace as European netizens, for example. The nature of their cyber citizenship reflects and is bounded by the nation state in which they are connected and is shaped by the foreign policy and security interests of that state.

This is an issue for serious international crimes (such as cyber terrorism), which by their nature require international cooperation. As cyber terrorists can often operate from not just one external location, but multiple, such activities are jurisdictionally complex, and investigation and prosecution are extremely difficult. Therefore, a holistic, harmonised, and transnational legal regime is required to deal with the extraterritorial nature of these crimes, yet such a supranational regime (without new forms of citizenship) will necessitate a controversial ceding of power away from individual states.[41] Policy scholars of both outer space and cyberspace must find a balance between respecting national sovereignty and practical security interests, satisfying the broadest possible range of stakeholders, and maintaining the integrity of both outer space and cyberspace as a global commons, for the benefit of humanity.

Having surveyed the current international legal framework concerning citizens in outer space and cyberspace, I will now draw on the typology previously outlined

---

[37]Kittichaisaree, Kriangsak. 2017. *Public International Law of Cyberspace.* New York: Springer.

[38]The General Data Protection Regulation (EU) 2016/679 (GDPR), Article 3.

[39]Yeli (n 6) 109.

[40]A portmanteau of 'internet' and 'citizen'.

[41]Yeli (n 6) 112.

and the ensuing discussion, and offer preliminary insights into a new, cosmopolitan citizenship for outer space and cyberspace.

2.5 New Modes of Citizenship for the Global Commons

2.5.1 Legal Cosmopolitanism

As mentioned, there is considerable debate in the fields of both outer space and cyberspace regarding the extension (or lack thereof) of state sovereignty into the global commons. But what if the 'global commons' aspect of outer space and cyberspace was extended to the international state system, in order to facilitate a consistent legal regime, that allowed permanent citizens of all three spheres to live and interact freely?

Legal cosmopolitanism as a theory emerged in the 90s, in response to globalisation and the increasing scope of global governance. Building on Kantian ideas of world citizenship,[42] legal cosmopolitanism is:

> committed to a concrete political ideal of a global order under which all persons have equivalent legal rights and duties, that is, are fellow citizens of a universal republic.[43]

Theorists of legal cosmopolitanism see the increasing interconnection of the international sphere as an encouraging development, as human concerns are ever more global, and should be considered on a planetary basis.[44] They suggest that as these developments continue apace, we should increasingly reconceptualise basic rights, legal obligations and political decision making at the global level when needed, and with it our concept of citizenship. This is not to argue for a world government, but rather a vertical redistribution of sovereignty, 'a kind of second-order decentralisation away from the now dominant level of the state',[45] with something akin to the EU's principle of subsidiarity, whereby decision making is delegated as much as possible to the lowest relevant level, but up to and including the planetary (or universal). In a cosmopolitan legal order, rights and duties are considered at the level at which they most effect the people subject to them.[46] Institutions like the European Court of Human Rights and the International Court of Justice are potential precursors to a legal cosmopolitan international regime, to the extent that their rulings are enforceable, and their competences remain strictly limited to the supranational or international level. The increasing prominence of

[42] Linklater (n 25).

[43] T Pogge, 'Cosmopolitanism and sovereignty' [1992] 103(1) Ethics 49.

[44] D Chandler, 'New rights for old? Cosmopolitan citizenship and the critique of state sovereignty' [2003] 51(2) Political Studies 332-349.

[45] T Pogge (n 46) 58.

[46] AS Sweet and C Ryan, A cosmopolitan legal order: Kant, constitutional justice, and the European Convention on Human Rights (Oxford University Press 2018).

two global commons in international law (outer space and cyberspace) are contentious when considered primarily through the paradigm of Westphalian sovereignty, but are consistent with the legal cosmopolitan perspective, and thus their existence encourages us to re-evaluate traditional modes. The planetary scope of cosmopolitan citizenship theory may constitute an escape from the impasse of state sovereignty, one that is consistent with the Outer Space Treaty and the globalised nature of cyberspace.

In practice therefore, permanent human existence in global commons necessitates the development of a new, cosmopolitan citizenship. This new form of citizenship, however, is contingent on a cosmopolitan redistribution of sovereignty across the international sphere, also. To this end, the international community should pursue to the greatest possible extent this reconfiguration, insofar as an increased presence of humans in the global commons necessitates it. This could potentially be done via 'gradual global institutional reform'[47] (e.g., a global forum; strengthening of international institutions; popular vote).

Some have argued that cosmopolitanism allows for a greater 'reflective equilibrium', in that citizens who are increasingly exposed to people from around the world develop a more holistic and well-rounded worldview. Similarly, exposure to new individuals in cyberspace can promote epistemological friction, 'cultural exchanges, economic cooperation, and collaborative security efforts'.[48] In the same vein, settlements in outer space could become a region to exchange ideas with people of different backgrounds and help promote a more planetary consciousness, as the spirit of solidarity and international co-operation on the ISS has shown. A cosmopolitan citizenship which acknowledges both as a global commons is beneficial not only from a legal standpoint, but also from a moral one, in terms of the development of humankind.

2.5.2 Global Netizenship

If we apply this lens of cosmopolitan subsidiarity to cyberspace, it is clear that it constitutes a multifaceted and complex legal arena, in which some issues may be national and many others global. In practice, the cosmopolitan approach would encourage us to think of citizens in cyberspace as acting as part of a planetary humanity ('global netizens'), who are responsible to the court that corresponds to the nature of the given issue, up to and including the global, and indeed: outer space. Additionally, they will be entitled to basic rights pertaining to their use of cyberspace (such as freedom of speech, freedom of assembly), and political determination within virtual spaces (analogous to the *virtual diasporas* previously discussed). As these netizens still exist in the physical world however, their

[47]T Pogge (n 46) 48.

[48]Yeli (n 6) 112.

Table 2.1 Modes of citizenship in global commons under different international legal systems

Sphere of activity	Mode of citizenship	
	Traditional Westphalian international system	Legal cosmopolitan international system
Planet Earth	National	Cosmopolitan
Cyberspace	Supplementary; contested	Cosmopolitan ('global netizenship')
Outer space	Extraterritorial; contested	Cosmopolitan ('extraterrestrial citizenship')

netizenship will be fundamentally *supplementary* to any national citizenships, until such a time as cosmopolitan citizenship is the prevailing international mode (see Table 2.1). In other words: while national citizenship takes precedence in the international system outside of cyberspace, netizens will have basic rights and responsibilities to a. their state of origin, and b. a global network, to the extent that their actions are relevant to these realms. The boundaries and various competences would be negotiated[49] between states, individuals and the international community, until such a time as a form of legal cosmopolitan citizenship and governance pertained in both cyberspace and physical space (Earth).

2.5.3 Extraterrestrial Citizenship

What would a cosmopolitan citizenship look like in practice for long-term (or permanent) residents of outer space settlements? If we apply the lens of cosmopolitan subsidiarity, we can see that everyday legal problems in such environments (interpersonal disputes, births, deaths, marriages) are for the most part beyond the international state system, and thus will have to be dealt with -to the greatest extent and whenever possible- by the residents of those settlements. This of course leaves out private property and real estate, which is prohibited by Article II of the Outer Space Treaty. It should, however, provide a framework for people and communities to begin to settle and live dignified and meaningful lives in outer space, without necessarily being considered colonists on the part of their nation of origin, or legally dependent on planet Earth. In practice, this would be a function of the settlement's real-world autonomy, but the principle would be to continually enhance the self-determination of the people in these settlements and consider them planetary or 'world' citizens. As with netizenship, however, a functional extraterrestrial citizenship is contingent on a legal cosmopolitan regime on Earth (see Table 2.1).

---

[49]Collyer (n 14) 56.

2.6 Conclusion

When the Apollo 11 astronauts emerged from their lander on July 20th, 1969, it was not an Earth flag or UN flag that they planted in the Sea of Tranquillity, but a national, American one.[50] Though perhaps contradicting the 'spirit' of the Outer Space Treaty, it was legal, as the astronauts were not making a formal conquest of the lunar surface, but rather a political statement: they were extraterritorial citizens of the United States, on a brief sojourn, and soon to return home. The next humans to visit the Moon may not be returning home so soon, and indeed within the next few decades we may begin to see permanent lunar residences, as well as corollaries in low-Earth orbit, cislunar space, and beyond. These new long-term settlements will present a multitude of legal challenges and contradictions for their denizens, as well as for lawyers on Earth, tasked with an ever more complex array of legal questions. Thus, scholars of space law and policy should begin to formulate a new model of citizenship in anticipation of these developments, one commensurate with the environment of space as a global commons, a terra nullius beyond the reach of traditional state sovereignty. As I have discussed, there are a multitude of similarities here with the field of cyberspace, where virtual citizens (or netizens) are already adapting traditional modes of citizenship to a global commons. In light of these aspects, I have argued that a cosmopolitan citizenship is the most consistent and appropriate for outer space and cyberspace, but that its functioning is dependent on a rethinking of citizenship on Earth.

Connor Hogan holds a bachelor's degree in politics, Philosophy and Economics (PPE) from Queen's University, Belfast (QUB), for which he completed a dissertation on public and private initiatives in the space industry. He is currently studying for a MSc in Politics at the School of Politics and International Relations (SPIRe) in University College Dublin (UCD) and is undertaking a thesis on space policy and law.

[50]Although a UN flag was briefly considered, it was unanimously decided by a survey of NASA's top administration that an American flag should be chosen. See: AM Platoff, 'Where no flag has gone before: political and technical aspects of placing a flag on the moon' [1994] 1(1) A Journal of Vexillology 3–16.

Chapter 3
On the Dangers of Enclosing the Intangible: Applying Pistor's "Code of Capital" Critique to "Space 3.0" and DLT from an Anti-monopoly Perspective

Maria Lucas-Rhimbassen

Invisible things are the only realities.

—Edgar Allan Poe

Abstract The purpose of this chapter is to determine whether the private space industry can escape jurisdiction by dematerializing the space resources commerce processes and assets thanks to cyber technology. Investigated strategies include dematerialization of space resources into intangible assets (such as intellectual property rights and financial assets) before transforming them into digital or cyber resources via tokenization to be part of commercial transactions through smart contracts on a distributed ledger technology (DLT). This chapter asserts that pure code, devoid of legal contract and human oracles, can indeed escape jurisdiction. However, if the smart contract is hybrid, contractual law applies and connecting factors have an increased chance to succeed in case of human arbitrators as oracles. Nonetheless, to prevent further privatization of the law, opacity and anti-competitive behavior, transparency measures must be implements at the contractual and arbitrators' level. The methodology behind this chapter's rationale is based on Pistor's "Code of Capital" which elaborates on the detrimental enclosure of knowledge and how it can be accelerated by decentralized technology, which in fact fuses monopolistic ambitions. Last, this line of thinking is juxtaposed on Israel's Space 3.0 categorization of the contemporary space context to extrapolate Pistor's thinking to the space economy.

M. Lucas-Rhimbassen (✉)
Chaire SIRIUS, IDETCOM, University of Toulouse 1 Capitole, Toulouse, France
e-mail: Maria.lucas-rhimbassen@ut-capitole.fr

A. Froehlich (ed.), *Outer Space and Cyber Space*, Studies in Space Policy 33,
https://doi.org/10.1007/978-3-030-80023-9_3

3.1 Introduction

The commerce of space resource is highly mediatized and is becoming mainstream. It is known as the next space gold rush, the next space race, the next frontier, etc. However, this feat is largely due to the private sector within spacefaring nations who convinced national regulators to pass legislation encouraging and incentivizing such commerce, while not "departing" from compliant interpretation of space law principles such as non-appropriation of celestial bodies. Given this commercialization of space activities context, this chapter aims to analyze the legal issues pertaining to commercializing space resources and raises the question as to whether there is a risk that the private sector escapes jurisdiction altogether in the near term, thanks to cyber strategies and further dematerialization efforts while relying on intangible assets such as intellectual property (patents, trade secrets, etc.), financial assets (e.g., deregulated financial derivatives), smart contracts and tokens (such as non-fungible tokens or NFTs). The main layout of the analysis on which this text is based is borrowing from Pistor's works—particularly on the "Code of Capital"—which asserts that intangible assets such as intellectual property enclose knowledge, cause behind the scenes consolidation and unfair competition (e.g., monopolies) and that this tendency is facilitated by decentralized distributed ledger technology. This line of thought enabled shaping the logic of this text, in alignment with Israel's categorization of the space sector into three main stages: Space 1.0 (treaty law); 2.0 (national legislation) and 3.0 (decentralized private sector).

Space 3.0 is defined by Israel as:

> Space Governance 3.0 will be **inter-operator**: **private law regimes** constructed from **contracts** between spacecraft operators (and spacecraft, in some cases) in which all **space actors, public and private, play on a level field.**[1] (emphasis added)

The purpose of this text is therefore to apply Pistor's arguments to the "level field" of Space 3.0 and to understand the inherent legal issues, identify potential loopholes and formulate recommendations to ensure compliance of the space commerce 3.0 with the ethical principles as enshrined in international space law.

3.2 Transforming Space Resources into Cyber Resources

With the advent of space resources commerce beckoning, with some of its concrete foundations already in the works, many key questions arise about the legal aspects involved with the different components thereof. This section outlines a brief

[1]Israel, B., quoted in Galvan, B. M., "Blockchain and Space Law 3.0: How smart contracts can help govern the moon and outer space: As nations and private firms venture farther into the galaxy, blockchain and distributed ledger technology could help enforce rules and resolve conflicts", Forkast, October 15, 2020, retrieved from: https://forkast.news/blockchain-space-law-smart-contracts-govern-moon-outer-space/ (Israel).

overview of what space resources would legally entail and identify a potential loophole enabling space resources to escape jurisdiction through their dematerialization (e.g., "tokenization") as intangible assets. According to the rationale elaborated upon *infra*, there is an inherent risk of enclosing the space economy through monopolization schemes in all opacity and a resulting need to anticipate such a scenario and reorient multistakeholder efforts towards engaging in a dialogue to deflect the enclosure and reflect on solutions out in the open.

The dematerialization currently most *en vogue* pertains to digitalized tokens made available on distributed ledger technology (DLT) such as the blockchain. A token is the digitalized version of an asset and can either be fungible or not fungible (NFT). Scatteia further defines asset tokenization:

> Asset Tokenization is the process of converting some form of an asset into a digital token that can be moved, recorded or stored on a blockchain system, where the asset can be manipulated as a digital token. Tokenizing an asset would enable one to manage **its value exchange based on the** contract written into the blockchain network. Tokens issued through the Asset Tokenization process are a special type called "asset-backed tokens" or "security tokens", which act as claims to the underlying assets. These tokens might represent any asset, including a song, a kilowatt-hour of solar energy, a square meter of real estate or a **square kilometer of an asteroid**. As an example, tokenization allows one to tokenize a property, thus enabling one to purchase only two square meters of a fifty square meter house (…) Among others Space Asset Tokenization Enabling a crypto token-based **ownership** of space assets including spacecrafts, satellites and potentially, **astronomical bodies such as asteroids**.[2] (emphasis added)

In the space sector, it is advanced that space assets (from space resources to geospatial data) can be "tokenized" and enter thus the cyber realm of financial transactions on the blockchain in exchange for cryptocurrency, as described by Scatteia, below:

> Potential **space resource utilization** approaches including **asteroid mining and regolith extraction** on the Moon could be **facilitated by the tokenization of assets**. Tokenizing space resources has a huge range of applications in the space mining industry, since blockchain provides a mechanism to register the physical location of space resources as digital tokens, and track their transactions, thus enabling for a transparent identification and management process. (emphasis added)

Although Scatteia lists "an asteroid" as a tokenizable space asset in the PwC report mentioned above, in the eyes of the international space law principle of

---

[2]Scatteia, L. et al., What are the Applications for the Space Industry? Concepts and Definitions. Blockchain. PwC Report, 2019. Retrieved from: https://www.pwc.fr/fr/assets/files/pdf/2019/03/fr-pwc-space-and-blockchain-2019.pdf. The report further elaborates at p. 4 on use cases and states that: "ConsenSys, a blockchain firm, recently acquired Planetary Resources, an asteroid mining company, in order potentially to initiate the application of this use case. It is possible that, in a few years, ConsenSys could manage the transactions of Planetary Resources using its Ethereum-based blockchain network, thus enabling a new wave of investment for space resource utilization by an entity, irrespective of its geographical location".

non-appropriation, as enshrined within the article II of the Outer Space Treaty,[3] this would qualify *prima facie* as non-compliant with treaty law which forbids national appropriation of celestial bodies. Nonetheless, there is room for debate as article II is broad and results already in divergent interpretations and therefore generates loopholes. Two questions arise. Firstly, whether "tokenizing" an asteroid would amount to national appropriation. Secondly, whether an asteroid is characterized, under treaty law, as a celestial body. The two questions remain, as of this writing, unanswered.

3.2.1 Tangible Space Resources

Compellingly, treaty law (or the *corpus juris spatialis*) is also silent on defining "space resources", which generates debate at the governance level in terms of establishing what could be considered as "commons" in outer space, and at the legal level in terms of determining the applicable property rights with respect to the "non appropriation" principle as per article II of the Outer Space Treaty. This legal void has left enough room to be filled in by contemporary national space legislation, such as in the case of the United States (US), Luxembourg, and the United Arab Emirates (UAE), however, at the price of international fragmentation. The respective States behind such legislation prioritize their own geopolitical interests which also include industrial growth, and therefore, they strive to meet the latter's demands which differ from country to country. Additionally, the industry plays an ever more important role in further defining non-binding guidelines and frameworks at the international level through interdisciplinary initiatives such as the Hague International Space Resources Governance Working Group (HISRGWG), rallying representatives from government, academia, civil society, and the private sector, with the mission to submit their "Building Blocks" to the United Nations Committee on Peaceful Uses of Outer Space (UNCOPUOS).

As of this writing, the most precise definition of space resources in US law can be found below:

> §51301. Definitions
>
> In this chapter:
>
> (1) **Asteroid resource.** - The term "asteroid resource" means a space resource found on or within a single asteroid.
>
> (2) **Space resource.** -
>
> (A) In general. - The term **"space resource" means an abiotic** resource in situ in outer space.

[3] Treaty on Principles Governing the Activities of States in the Exploration and Use of Outer Space, including the Moon and Other Celestial Bodies, 1967, available online at: https://www.unoosa.org/oosa/en/ourwork/spacelaw/treaties/introouterspacetreaty.html.

(B) Inclusions. -The term "space resource" includes **water and minerals**.[4] (emphasis added)

However, at the international level, the HISRGWG went a step further in their own definitions submitted to UNCOPUOS to build on the momentum of US law towards space commerce:

2. Definition of key terms

2.1 **Space resource**: an **extractable and/or recoverable abiotic resource in situ in outer space**[5];

2.2 Utilization of space resources: the recovery of space resources and the extraction of **raw mineral or volatile materials** therefrom.

2.3 **Space resource activity**: an activity conducted in outer space for the purpose of searching for space resources, the recovery of those resources and the extraction of raw mineral or volatile materials therefrom, including the construction and operation of associated extraction, recovery, processing, and transportation systems.

2.4 **Space object**: an object launched into outer space from Earth, including component parts thereof as well as its launch vehicle and parts thereof.

2.5 Space-made product: a product made in outer space wholly or partially from space resources; (…).[6] (emphasis added)

Besides the US Space Act of 2015 and the HISRWG, space resources are mentioned in the Luxembourgian Space Resources Act of 2017, regulating their appropriation and commercialization, despite, however, further defining them.[7]

3.2.2 *Property Rights and Antitrust Issues: Thoughts of Enclosing Outer Space*

Interestingly,—although the French Space Operations Act of 2008 is the first national legislation centered on "space objects"—the HISRGWG filled, through

[4]51 USC 51301: Definitions (From Title 51-NATIONAL AND COMMERCIAL SPACE PROGRAMS Subtitle V-Programs Targeting Commercial Opportunities, CHAPTER 513-SPACE RESOURCE COMMERCIAL EXPLORATION AND UTILIZATION), retrieved from: https://uscode.house.gov/view.xhtml?req=granuleid:USC-prelim-title51-section51301&num=0&edition=prelim#sourcecredit.

[5]The section however reiterated the inclusions made by US law but excludes precisely certain other resources: "According to the understanding of the Working Group, this includes mineral and volatile materials, including water, but excludes (a) satellite orbits; (b) radio spectrum; and (c) energy from the sun except when collected from unique and scarce locations".

[6]BUILDING BLOCKS FOR THE DEVELOPMENT OF AN INTERNATIONAL FRAMEWORK ON SPACE RESOURCE ACTIVITIES, November 2019, retrieved from: https://www.universiteitleiden.nl/binaries/content/assets/rechtsgeleerdheid/instituut-voor-publiekrecht/lucht–en-ruimterecht/space-resources/bb-thissrwg–cover.pdf.

[7]Loi du 20 juillet 2017 sur l'exploration et l'utilisation des ressources de l'espace, retrieved from: http://data.legilux.public.lu/file/eli-etat-leg-loi-2017-07-20-a674-jo-fr-pdf.pdf.

soft law, the gap concerning the definition of a "space object", which besides doctrine,[8] if left unfilled and results in problematic consequences for determining specific property rights with respect to space resources systems and units when there is no scientific nor legal consensus as to what a "celestial body" or a "space object" consist of exactly. Legal requirements, such as "movable" or "immovable" goods, in terms of property rights are difficult to ascertain in this context,[9] which incidentally make the situation even more complicated to regulate. Nonetheless, given national legislation (e.g., the US and Luxembourgian case), backing international endeavors such as the HISRWG, which in turn enable nascent bilateral efforts paving the way to imminent space commerce such as NASA's Artemis Accords of 2020,[10] the international space law regime of non-appropriation seems to be circumvented in that "celestial bodies" as such are not to be appropriated, but space resources commercially extracted therefrom can be. Therefore, the applicable forum for space resources is national law, *inter alia* in the light of article VI of the Outer Space Treaty which provides that:

> **States Parties to the Treaty shall bear international responsibility for national activities in outer space**, including the Moon and other celestial bodies, whether such **activities are carried on by governmental agencies or by non-governmental entities**, and for **assuring that national activities are carried out in conformity with the provisions set forth in the present Treaty**. The activities of non-governmental entities in outer space, including the Moon and other celestial bodies, shall require **authorization and continuing supervision by the appropriate State** Party to the Treaty. When activities are carried on in outer space, including the Moon and other celestial bodies, by an international organization, responsibility for compliance with this Treaty shall be borne both by the international organization and by the States Parties to the Treaty participating in such organization. (emphasis added)

Article VI provides indeed that States are responsible for the space activities undertaken by their national non-governmental entities, inferring thus that non-governmental actors are subjected to national space legislation. The US Space Act of 2015 confirms this at section 402:

> A U.S. citizen engaged **in commercial recovery** of an **asteroid resource** or a space resource shall be **entitled to any asteroid resource or space resource obtained**, including **to possess, own, transport, use, and sell it** according to applicable law, including U.S. international obligations.[11] (emphasis added)

---

[8]Jakhu, R., Pelton, J., (eds), "Global Space Governance: An International Study", Space and Society Series, Springer, 2017., p.396. "Space objects" could include space mining facilities and equipment.

[9]Pop, V., "Who Owns the Moon: Extraterrestrial Aspects of Land and Mineral Resources Ownership", Springer, 2009. The question applies both to space resources and celestial bodies (e.g., is an asteroid a space resource or a celestial body? What is the minimal size of a celestial body? How does the characteristic of "immovable" meet legal requirements in space? etc.).

[10]NASA Artemis Accords of 2020, retrieved from: https://www.nasa.gov/specials/artemis-accords/img/Artemis-Accords-signed-13Oct2020.pdf.

[11]TITLE IV—SPACE RESOURCE EXPLORATION AND UTILIZATION (Space Resource Exploration and Utilization Act of 2015).

The Luxembourgian Space Resources Act of 2017 equally emphasizes throughout the legal document the need of non-governmental actors to be fully authorized by the State relating to the acquisition of space resources (activity referred to as "*agrément*"). On top of that, property rights law is also subjected to national law as there is no international property right framework.

Israel postulates that national space legislation ("Space 2.0") is compliant *de facto* with international space law ("Space 1.0) in that its core mission is to transplant international space law into the national jurisdiction. Space 1.0 includes the national obligation of States to supervise the space activities of their nationals, and this would imply an *ab initio* compliance to treaty law from the part of the authorized national entities, because of the national authorization itself. In fact, Israel's reasoning can be explained through both a pyramidal and parallelistic logics (Fig. 3.1.).

These illustrations aim to demonstrate Israel's thinking in the sense that, on the one hand, Space 3.0 cannot exist without the other levels, but, simultaneously, on the other hand, the three levels exist in parallel, with Space 2.0 as the bridge. Within the boundaries of this thinking, Space 3.0 cannot afford not to comply with Space 1.0 which is implemented through Space 2.0 under the form of authorization impacting 3.0. This reasoning legitimizes the American and Luxembourgian interpretation of treaty law according to which appropriation of space resources would be permitted. Notably, section 402 of the US Space Act of 2015 on space resources reiterates the compliance with treaty law "according to applicable law, including U.S. international obligations".

Furthermore, given the commercialization of the space sector and the emerging critical role to be played by the private sector at the level 3.0, the days of purely international law in the space context might be numbered due to the entrance of a more transnational commercial law, which is to evolve through opacity, stealth and elusiveness if not closely monitored. The sections *infra* delve deeper into these

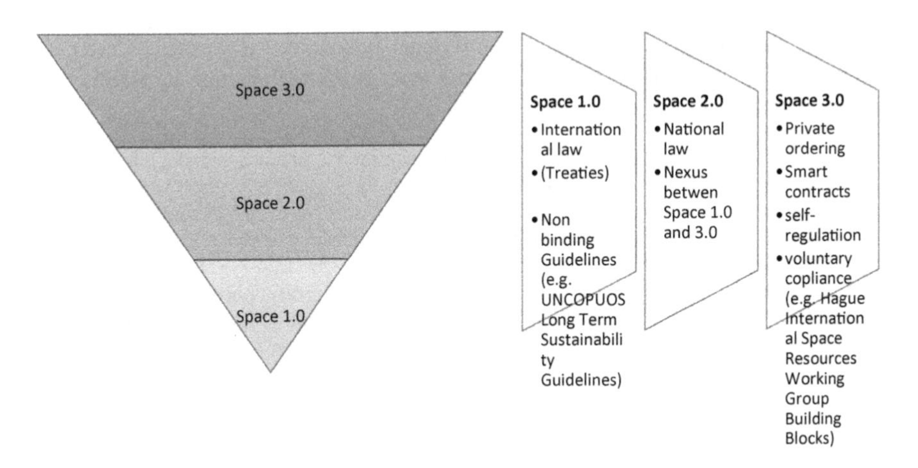

Fig. 3.1. Pyramidal and parallelistic illustrations of Israel's Space 3.0 rationale

hazardous prospects following multiple strategies of dematerializing property into intangible assets, such as tokenized intellectual property rights based on space resources and financial assets such as increasingly deregulated derivatives. The relevance of addressing intangibility within the space situation is that the more an asset gains in intangibility, the more it becomes diluted and decentralized and the more chances it must escape any particular jurisdiction's control, which seems to add to the commercial motivation behind today's budding space race. The first dematerialization scheme to be analyzed below is intellectual property, which most importantly, besides raising jurisdiction queries, ushers in antitrust issues entangled in enclosure.

3.2.3 Space TRIPS: Intellectual Property (IP)

IP regime in space is far from clear. In fact, it is so unclear since the World Intellectual Property Organization (WIPO) 2004 report on IP and space activities[12]—requested by the Organization for Economic Co-operation and Development (OECD), following its "Futures Project on the Commercialization of Space and the Development of Space Infrastructure"—there are no major advances in regulation. The report stresses on the observation that this legal uncertainty would be detrimental to competition, implying thus antitrust issues:

> The importance of establishing a legal regime that effectively protects intellectual property in space cannot be overemphasized. Lack of legal certainty will influence the advancement of space research and international cooperation. Because of the large investments involved in space activities, a **legal framework that assures a fair and competitive environment is necessary to encourage the private sector's participation** in this field. Limited exclusive rights conferred by intellectual property protection would bring competitive benefits to right holders either by concluding a licensing agreement or by excluding competitors from using a given technology.[13] (emphasis added)

The report further concludes that this legal uncertainty is to cause "complex segmentation" of quasi-territorial IP and that multiple bilateral or multilateral agreements are to be conducted in each case:

> (...) **in the absence of legal certainty** as to how the territorial jurisdiction under intellectual property law could apply to extraterritorial activities on a spacecraft which is subject to nationality jurisdiction, in practice, registered space objects are treated as **quasi-territory** for **the purposes of intellectual property** under a number of international agreements concluded with respect to international space projects. This leads to a **patchwork** of national intellectual property laws each of which could only be applicable on a relevant registered object. It means that, in the case of international cooperation activities, a

[12] WIPO, "Intellectual Property and Space Activities", Issue paper prepared by the International Bureau of WIPO, April, 2004, retrieved from: https://www.wipo.int/export/sites/www/patent-law/en/developments/pdf/ip_space.pdf (WIPO).

[13] Ibid., p. 5, paragraph 21.

> **complex segmentation** of the international space station, or any other future international platform, **cannot be avoided**. Further, the **lack of a global agreement** leads to the situation that an **agreement would have to be concluded** among the parties in **each single case** of international cooperation. (emphasis added).

Smith equally uses the term "patchwork" while recalling that it took over a decade to negotiate IP rights applicable on the International Space Station (ISS) and describing results unfolding into a patchwork of modularity.[14] He equally states that the European Community Patent ("Unitary Patent System") still in progress as of this writing[15] is reminiscent of this patchwork while the proposed article 2 would extend jurisdiction to outer space.[16] He also deplored the fact that with regards to the space sector, at the international level, the UN efforts during the UNISPACE III conference, in 1990, have not crystallized in concrete action in three decades[17]; that the WIPO has dodged the question by stating that a space patent is like any other patent; and that the World Trade Organization (WTO)'s Agreement on Trade-Related Aspects of Intellectual Property Rights (TRIPS) "ignored the problem".

While international efforts seem to be stalling, the vacuum is only filled by US national legislation, with the US Patents in Space Act of 1990 which is the "only explicit provision establishing a link between the three key elements: inventions, jurisdiction and territory"[18] and which establishes quasi-territoriality unless otherwise specified by an international agreement:

> Section 105 of 35 U.S.C. (Inventions in outer space) reads as follows:
>
> (a) **Any invention made, used, or sold in outer space on a space object or component thereof under the jurisdiction or control of the United States** shall **be considered** to be made, used or sold within the **United States** for the purposes of this title, **except** with respect to any space object or component thereof that is specifically identified and **otherwise provided for by an international agreement** to which the United States is a party, or with respect to any space object or component thereof that is carried on the registry

---

[14]Smith in Myers, G., "Intellectual Property Resources in and for Space: The Practitioner's Experience", University of Missouri School of Law Scholarship Repository Faculty Publications, 2006, retrieved from: https://scholarship.law.missouri.edu/cgi/viewcontent.cgi?article=1291&context=facpubs, (Myers, 2006), p. 411.

[15]The Unitary Patent System is expected to apply starting 2022. See European Patent Office (EPO) at: https://www.epo.org/law-practice/unitary/unitary-patent/start.html. Article 2 provides that : "This Regulation shall apply to inventions created or used in outer space, including on celestial bodies or on spacecraft, which are under the jurisdiction and control of one or more Member States in accordance with international law".

[16]Proposal for a Council Regulation on the Community patent, (2000/C 337 E/45), (Text with EEA relevance).

COM(2000) 412 final 2000/0177(CNS), (Submitted by the Commission on 1 August 2000), Official Journal of the European Communities 28.11.2000, retrieved from: https://eur-lex.europa.eu/legal-content/EN/TXT/PDF/?uri=CELEX:52000PC0412&from=EN.

[17]More on the UNISPACE III in Benkö, M., et al., "Space Law at UNISPACE III: Achievements and Perspectives", DLR, January 2000, retrieved from: https://www.researchgate.net/publication/224784866.

[18]WIPO, note 13, *supra*, p. 11, paragraph 43, citing Section 105 of Title 35 of the U.S.C.228.

of a **foreign state** in accordance with the Convention on Registration of Objects Launched into Outer Space.

(b) Any invention made, used or sold **in outer space on a space object or component thereof that is carried out on the registry of a foreign state** in accordance with the Convention on Registration of Objects Launched into Outer Space, shall be considered to be **made**, used or sold within the **United States** for the purposes of this **title if specifically so agreed in an international agreement** between the United States and the state of registry. (emphasis added)

In the space sector (e.g., space mining industry), without such agreements, there is a risk that multiple jurisdictions exist in parallel and that same processes benefit from different IP rights under these jurisdictions, submitted by competing actors, causing thus a conflictual IP conundrum between rival commercial entities:

If a mining lease used to prevent third parties from mining a deposit cannot be used for celestial mining, it is likely that greater **emphasis** will be placed on technology that can **mine the resource faster than a competitor.** It naturally follows that **significant IP will arise in the development of such technology**. If a private entity develops a machine to allow them to mine a resource on a celestial body faster than their competitors and they only **obtain patent protection for the IP of the machine in the same jurisdiction** in which the machine is registered for the purpose of the Treaty, a **third party may be able to copy** the machine and IP and mine the same deposit without worry of infringement if the third party **registers their machine in a jurisdiction in which the entity does not have patent protection**. Such a situation shows why determining the jurisdiction(s) in which a competitor may register their technology for the purpose of the Treaty is important for IP enforcement.[19] (emphasis added)

This puzzle of "rivalrous" IP jurisdictions can amount to a situation comparable with the "flags of convenience" complex problem whereby commercial actors register their activities under a more advantageous jurisdiction.[20] In short, Avedutto summarizes well this situation by stating that "territoriality does not fit space activities"[21]:

The **commercial dimension of outer space**, including the **assertion of intellectual property rights**, cannot rely on territorial definitions that only work when supported by the consensus of the concerned parties. **Territoriality is also incompatible with the non-appropriation** principle of the Outer Space Treaty, prohibiting sovereign claims or occupation: absent clear boundaries over certain portions of space, **recourse to legal fictions and extraterritorial laws will become more widespread, fueling inconsistencies**

[19]Paterson, S., et al., "The role of intellectual property in space", Spacetech, July 31, 2018, retrieved from: https://www.spacetechasia.com/the-role-of-intellectual-property-in-space/.

[20]Taghdiri, A, "Flags of Convenience and the Commercial Space Flight Industry: The Inadequacy of Current International Law to Address the Opportune Registration of Space Vehicles in Flag States", 19 B.U. J. SCI. & TECH. L. (2013) p. 405, 407; Ro, T. U. et al., "Patent Infringement in Outer Space in Light of 35 U.S.C. § 105: Following the White Rabbit Down the Rabbit Loophole", 17 B.U. J. SCI. & TECH. L. (2011), p. 202, 212–13.

[21]Avvedutto, R., "Past, Present, and Future of Intellectual Property in Space: Old Answers to New Questions", Washington International Law Journal, Volume 29 Number 1, 12-23-2019, retrieved from: https://digitalcommons.law.uw.edu/cgi/viewcontent.cgi?article=1829&context=wilj., at p. 237.

and frictions. (…) The absence of a defined authority in space creates a **vacuum of power and an incentive for unaccountable or unsupervised activities** (…).[22] (emphasis added)

3.2.4 *IP for Intellectual Parceling?*

IP, as explained supra, falls under domestic jurisdiction[23] and comprises elements such as patents, trademarks, property rights and others (e.g., trade secrets or business processes).[24] The US Patent Act of 1790, in its currently valid version, provides that:

> (...) whoever **invents or discovers any new and useful process, machine, manufacture, or composition of matter**, or any new and useful **improvement** thereof, may **obtain a patent** therefor, subject to the conditions and requirements of this title.[25] (emphasis added)

This Act grants a *de facto* monopoly power temporarily to the patent owner by precluding others from using the same processes protected by the patent for a standard period to secure competitive advantage. However, Pistor explains that the ultimate beneficiary of intellectual property such as copyrights, are corporate shareholders:

> The justification for creating **these temporary monopolies** is to **incentivize the inventor** or artist by allowing them to fully capture the monetary value of their creativity for fear would otherwise cease activities that might be of tremendous social value. Yet, human creativity has been driven over the millennia by motives other than monetary gains. Even with a comprehensive system of intellectual property rights in place, most authors, composers, and inventors receive only a tiny return for their creativity. The **ultimate beneficiaries of the legal monopolies that intellectual property rights create are corporations that extract returns from the patents for the financial benefits of their shareholders.**[26] (emphasis added)

According to Pistor,[27] quoting Pagano,[28] this monopolization of knowledge, pushed by corporate interests, contributes to a "secular stagnation" in that the "enclosure of knowledge is responsible for the decline in viable investment opportunities and has led to an investment famine" caused by a paradoxically shrinking number of globally expanding mega-corporations, which incidentally reduces knowledge diversity because of consolidation. One might argue that the

---

[22]Ibid, at p. 246.

[23]WIPO, *supra,* note 13, at p.10, paragraph 40.

[24]Pistor, K., The Code of Capital: How the Law Creates Wealth and Inequality, Princeton University Press, 2019 (Pistor) at p. 116.

[25]35 U.S. Code § 101—Inventions patentable, (July 19, 1952, ch. 950, 66 Stat. 797.), retrieved from: https://www.law.cornell.edu/uscode/text/35/101.

[26]Pistor, *supra,* note 25, p. 114.

[27]Pistor, *supra,* note 25, p. 117.

[28]Pagano, The Crisis of Intellectual Monopoly Capitalism, p. 1419.

next target for rampant IP is the space sector and the sections below elaborate on the space resources processes to be patented and transformed into intangible assets, diluted within—if not beyond—transnational legal boundaries.

3.2.5 Welcome to the Space Commerce of Tomorrow

The corporate world is not the exclusively responsible of the above-mentioned enclosure. It is indeed backed by States[29] which consequently became locked into an inescapable mesh of patent globalization dynamic initiated by the developed States at the detriment of the others. Pistor argues that IP harmonization efforts are reflecting extraterritorial ambitions in terms of imposing a "Western" culture upon the knowledge commons and failure to adhere to this coercive trend results in economic sanctions,[30] oftentimes "petitioned" by the private sector.[31] Besides the first international IP treaty, the 1883 Paris Convention for the Protection of Industrial Property,[32] focusing on reciprocity, which led to the creation of the World Intellectual Property Organization (WIPO) in 1970—with 193 member States as of this writing—the most significant international effort towards IP harmonization is the World Trade Organization (WTO)'s Agreement on Trade-Related Aspects of Intellectual Property Rights (TRIPS) which tend to focus on coercive enforcement. According to Pistor, TRIPS:

> (...) created major **carve-outs** from the free trade regime for **monopolies** under the **label of intellectual property rights**. TRIPS gave the technologically more advanced companies of the global North the option to **enclose their know-how** and thereby **remove free access** to it by **potential competitors** in less advanced countries.[33] (emphasis added)

This rationale foretells a consolidated scenario which if extrapolated to the space sector, then could facilitate a wide span of IP industrial monopolies beyond the ultimate frontier and further deprive humanity of knowledge diversity as in Pistor's words:

> By endorsing a **singular approach** based on the **business** interests in the most advanced economy, the **world missed a critical opportunity to create an intellectual property rights regime** of meaningful **diversity** and critically, to **preserve** at least parts of the **global commons in knowledge.**[34] (emphasis added)

The WIPO report concurs at least in that non-commercial research should not be captured by existing patents and that this concerns the space sector in particular:

---

[29]Pistor, *supra,* note 25, p. 115.

[30]Pistor, *supra,* note 25, p. 122.

[31]Pistor, *supra,* note 25, p. 121, on the new Trade Act of 1974, 19 USC, Chapter 12, Sec 301.

[32]The now WIPO, at: https://www.wipo.int/treaties/en/ip/paris/.

[33]Pistor, *supra,* note 25, p. 123.

[34]Pistor, *supra,* note 25, p. 126.

> It should be noted that, in the field of patents, many national laws provide that the rights conferred by a patent **should not extend to the use of patented subject matter for the purposes of non-commercial experimentation and research**. What constitutes "non-commercial" activities and the definition of the term "experimentation and research" differ from one country to another. The underlying objective of such an exception, however, is to provide a **balance between the patentee** who obtains **exclusive rights and third parties who wish to test the reproducibility and usefulness of the patented invention** or who wish to further develop the technology by way of experimenting with the patented subject matter. **Since many activities in outer space could be characterized as experimental and research activities, certain uses of patented technology in outer space may fall under the experiment and research exemption**, depending on the scope of the experiment and research exemption clause under the national law.[35] (emphasis added)

Finally, the report provides that questions arise as to whether IP enforcement conflicts with fundamental Outer Space Treaty principles (e.g., access to knowledge and information derived from space activities). Article I of the Outer Space Treaty stipulates indeed that:

> The **exploration and use** of outer space, including the moon and other celestial bodies, shall be carried out for the **benefit and in the interests of all countries**, irrespective of their degree of economic or scientific development, and shall be the **province of all mankind**.
>
> Outer space, including the moon and other celestial bodies, shall **be free for exploration and use by all States without discrimination of any kind**, on a basis of **equality** and in accordance with international law, and there shall **be free access to all areas** of celestial bodies.
>
> There shall be **freedom of scientific investigation** in outer space, including the moon and other celestial bodies, and **States shall facilitate and encourage international co-operation in such investigation**. (emphasis added)

This dilemma concerns the freedom of exploration and use of outer space versus and the possibility of excluding others from accessing outer space by way of obtaining IP rights (…).[36]

Such scenario has been identified by Pompidou in the case of patenting "intelligent orbits" which could contribute to limiting access to defined space areas, through other means than preventing harmful interference:

> The **protection of intellectual property** may have proved a **threat** to the development of **subsequent research**, especially in the case of intelligent orbits for which the necessary technology has been patented. These orbits are elliptical and quasi-geostationary but located outside the equatorial zone. This means that they do not require limited frequencies. The **consequence** of a patent registration is that people who would wish to **use these orbits for commercial or research** purposes will have to **pay royalties**. This practice has the effect of **limiting access to space and is hard to justify**.[37] (emphasis added)

The risk of further engulfing space resource into the IP realm consists in either modifying a space resource *per se*, as mentioned *supra* by the US law, or patenting

---

[35]WIPO, *supra*, note 14, paragraph 61.

[36]WIPO, *supra*, note 13, paragraph 66.

[37]Pompidou, A., "The Ethics of Space Policy", UNESCO Report, 2000, retrieved from: https://unesdoc.unesco.org/ark:/48223/pf0000120681, (Pompidou), p. 20.

the process of transforming the resources into other "baskets"[38] of space commodities[39]: processed goods, space-based services, contractual rights, and financial rights (such as derivatives). Through a combination of patents and trade secrets, each of these categories can generate potentially artificial monopolies downplaying a beneficial competitive environment. One of these processes can be the transformation of the said resources into intangible cyber resources *via* tokenization. It is not clear whether patenting technology pertaining to the usage of entire orbits contravenes to principle of non-appropriation, which is why there is a possibility that until proven otherwise, patent rights issued with regards to the technology determining intelligent orbits (e.g., modified trajectories) could be tokenized in the meantime, raising subsequently a bundle of policy and security questions.

3.2.6 Cyber Space Resources: Darker than Black Holes

The section above depicts the situation where space resources or technology related to space areas such as orbits can profit from dematerialization such as tokenizing respective IP rights (themselves intangible assets). Tokenizing the intangible, especially in an otherwise heavily regulated sector like space, might indeed result in a problematic outcome in terms of governance, policy, security, safety, and sustainability. It could also facilitate opaque strategies amounting to unfair competition such as concerted practice, monopolization, and collusion since there is no regulation pertaining to token acquisition, which can also be subdivided into multiple parts of that same token, be it fungible or not. Last, acquisition can come from the part of non-governmental entities (e.g., corporations), but also from various governmental actors, raising thus geopolitical concern. Although DLT advocates complete transparency, traceability in the cyber domain proves increasingly more challenging.

Cyber resources are defined as:

> (...) an information resource which creates, stores, processes, manages, transmits, or disposes of information **in electronic form,** and which can be accessed via a network or using networking methods.[40] (emphasis added)

Extrapolated to the space sector as space cyber resources, they could further profit from a nexus comprised of several layers dodging jurisdiction, more or less successfully, culminating in their placement into a decentralized DLT technology, as tradeable financial assets (e.g. derivatives, which are themselves subject to

---

[38]Cahan, B., "Space Commodities Panel", NewSpace 2017, video available online at: https://www.youtube.com/watch?v=EWEtBIxREsU.

[39]Ahadi, B., et al., "Space Resources Commodities Exchange", New Space, Vol. 8, No. 2, 12 June 2020, paper available online at: https://doi.org/10.1089/space.2019.0039.

[40]Cyber resources as defined in the Glossary of the National Institute for Standards and Technology, US Department of Commerce, retrieved from: https://csrc.nist.gov/glossary/term/cyber_resource.

deregulation following the Modernization Act of 2000[41]), which could be protected by monopolized IP market segments and protocols within the given platform and no regulator could intervene to sanction anti-competitive behavior enforced by automated mechanisms such as "smart contracts". Undeniably, it can indeed prove difficult to enforce national space legislation or international initiatives such as the HISRWG's Building Blocks to space resources once they become intangible as hereby illustrated. For this reason, the space resources definition's scope must imperatively be broadened to include intangibles to prevent downstream mayhem, first and foremost when regulating the cyber domain is mined with legal blurring hurdles and technical challenges. At this point, the analysis does not even begin to scrutinize issues surrounding controversial blockchain-related instruments such as cryptocurrency, which unfolds in substantial debate which could mature into more obscurity, no pun intended.

3.2.7 Loophole 3.0: Contracting for the Great Escape through Forum Shopping

As mentioned supra multiple times, intangible financial assets are beyond sovereign authority. In Pistor's words:

> When it comes to property rights, however, most States still insist on their legal sovereignty and impose domestic law on assets that are located within their own territory (…) **but territorial control is of little use for assets that lack physical form or location**; for **tradeable financial assets**, other criteria had to be found to determine whose law should govern them – and ideally criteria that would point to one and same legal system (…).[42] (emphasis added)

[41]US Commodities Futures Modernization Act which reduced regulation over speculation. Act available at H.R.5660—106th Congress (1999–2000), retrieved from: https://www.congress.gov/bill/106th-congress/house-bill/5660/text. Stout stated that "The CFMA not only declared financial derivatives exempt from CFTC or SEC oversight, it also declared all financial derivatives legally enforceable. The CFMA thus eliminated, in one fell swoop, a legal constraint on derivatives speculation that dated back not just decades, but centuries. It was this change in the law—not some flash of genius on Wall Street—that created today's $600 trillion financial derivatives market." See Stout, L., "Why re-regulating derivatives can prevent another disaster", Harvard Law School Forum on Corporate Governance & Financial Regulation, on Tuesday, July 21, 2009, retrieved from: https://corpgov.law.harvard.edu/2009/07/21/how-deregulating-derivatives-led-to-disaster/. Stout explains that: "The first "financial" derivatives, in the form of stock options, became common in the 1800s. The 1990s saw an explosion in other types of derivatives contracts, including bets on interest rates (interest rate swaps), credit ratings (credit default swaps), and even weather derivatives. By 2008, the notional value of the derivatives market—that is, the size of the outstanding bets as measured by the value of the things being bet upon—was estimated at $600 trillion, amounting to about $100,000 in derivative bets for every man, woman, and child on the planet. (…) This sudden development of an enormous market in financial derivative contracts was not the result of some new idea or "innovation." Rather, it was a consequence of the steady deregulation of financial derivatives trading".

[42]Pistor, *supra*, note 25, p. 135.

This one and the same legal system comes under the form of internationally standardized contracts framing financial assets and applicable conflict-of-law rules.[43] Subsequently, in 1985, financial derivatives[44] contracts were harmonized by the influential International Swaps and Derivatives Association (ISDA), whose main purpose is to facilitate scalability of financial assets on a globalized market[45] by and large through lobbying means. Last, financial derivatives benefited from United Kingdom's Financial Services Act of 1986, further deregulation in 2000 by way of the Modernization Act which leaves considerable room for legal creativity in terms of contractual architecture. However, this trend of deregulation can take such "creative" architecture to a whole new level on the account of new technology and algorithms equivalent to self-executory "smart contracts"[46] which, according to the blog of the Faculty of Law at the Oxford University, "many authors claim that 'code is law'[47] since smart contracts, especially when stored on and enforced *via* DLT, will make traditional contract law obsolete[48]" if smart contracts are considered not legal contracts but "merely a piece of software or program code that controls, monitors, or documents the execution of a contract that has been concluded elsewhere". Smart contracts, as Ruhl argues, need a "legal system as a normative point of reference",[49] determined by the rules of private international law.[50] In this case, in Europe, Rome I Regulation[51] would apply if this software came with a legally binding contract. Rome Regulation I permits either contractual

---

[43]Pistor, *supra*, note 25, p. 136 citing the works of the Hague Conference on Private International Law (HCCH) established in 1893.

[44]Financial derivatives are defined by Investopedia as "A derivative is a financial security with a value that is reliant upon or derived from, an underlying asset or group of assets—a benchmark. The derivative itself is a contract between two or more parties, and the derivative derives its price from fluctuations in the underlying asset. The most common underlying assets for derivatives are stocks, bonds, commodities, currencies, interest rates, and market indexes. These assets are commonly purchased through brokerages", retrieved from: https://www.investopedia.com/terms/d/derivative.asp.

[45]Pistor, *supra*, note 25, p. 145.

[46]Pistor, *supra*, note 25, p. 187.

[47]Lessig, L., "Code and other Laws of Cyberspace", Basic Books, 1999; Lessig, L., "Code is Law: On Liberty in Cyberspace", Harvard Magazine, January/February 2000, available at: https://harvardmagazine.com/2000/01/code-is-law-html.

[48]Ruhl, G., The Law Applicable to Smart Contracts, or Much Ado About Nothing?, 23 Jan. 2019, Oxford Business Law Blog (OBLB), retrieved from: https://www.law.ox.ac.uk/business-law-blog/blog/2019/01/law-applicable-smart-contracts-or-much-ado-about-nothing (Ruhl 2019).

[49]Ruhl, G., Smart (legal) contracts, or: Which (contract) law for smart contracts? forthcoming, in: Benedetta Cappiello & Gherardo Carullo (eds.), Blockchain, Law and Governance, Springer, 2020, retrieved from: https://papers.ssrn.com/sol3/papers.cfm?abstract_id=3552004 (Ruhl 2020), p. 4.

[50]Ibid., p. 5.

[51]"REGULATION (EC) No 593/2008 OF THE EUROPEAN PARLIAMENT AND OF THE COUNCIL of 17 June 2008 on the law applicable to contractual obligations (Rome I), retrieved from: https://eur-lex.europa.eu/LexUriServ/LexUriServ.do?uri=OJ:L:2008:177:0006:0016:EN:PDF.

choice of law by the parties and if there is none selected, because of the potential anonymous origin of smart transactions, Rome Regulation I assigns the applicable jurisdiction according to connecting factors to actors involved (known as "nodes"[52]), providing that transparency measures permit such tracing.

3.2.8 A Genesis of Oracles: Cyber Resilience

Ledgers are not infallible. Bugs, mistakes, or bad faith (e.g., hacking, corrupt data, etc.) can get the best of the entire infrastructure and cause the system's implosion and failure. Adaptation is necessary to repair and foresee future threats and build resilience. The community of interest pinpoint to the need of external beacons acting as neutral points of reference enabling readjustment which are known as "oracles"[53] which can retrieve "off-chain"[54] data and "feed a smart contract with e.g., benchmark prices (…) but they can also request a decision from an external arbiter".[55] Nevertheless, such oracles can be manipulative and serve a specific set of agendas conflicting with the infrastructure in question.[56]

As per Ricardian or hybrid contracts, space smart contracts advocates, such as Israel, posit that "having human arbitrators in the oracle loop" might bring the best of both formats together: automatization and enforceability.[57]

Space smart contracts advocates, such as Israel, emphasize their relevance as opposed to traditional contracting, besides being time-consuming, triggers frictions at different levels:

> In contract negotiations between nationals of different states, agreement on which state's laws will govern the interpretation of the contract, which state's courts will have jurisdiction over disputes, and which language will be authoritative in interpreting the contract's terms, can **consume as much time as negotiations over the content** of those terms. This **transactional friction** increases with the number of nationalities involved. Fixing these **jurisdictional terms at the time the original parties enter the contract makes it less likely that new satellite operators will opt-in to the contractual regime at a later date**. And the **high cost of enforcing a transnational contract** makes the **threat of sanctions for non-compliance less credible**, diminishing the deterrence value of the contractual regime as operators weigh the costs and benefits of compliance.[58] (emphasis added)

---

[52]Ruhl (2020), *supra*, note 50, p. 6.

[53]Pistor, *supra,* note 25, p. 189.

[54]Levi, S. D., et al., "An Introduction to Smart Contracts and Their Potential and Inherent Limitations", Harvard Law School Forum on Corporate Governance, Saturday, May 26, 2018, retrieved from: https://corpgov.law.harvard.edu/2018/05/26/an-introduction-to-smart-contracts-and-their-potential-and-inherent-limitations/.

[55]Pistor, *supra,* note 25, p. 189.

[56]Pistor, *supra,* note 25, p. 190.

[57]Israel, *supra*, note 2.

[58]Israel, B. R., "Space Governance 3.0", 48 Ga. J. Int'l & Comp. L. (2020). Available at: https://digitalcommons.law.uga.edu/gjicl/vol48/iss3/7, (Israel, 2020), p. 722.

Israel argues that smart contracts will solve these hurdles, however he does not directly raise cybersecurity issues which could undermine the whole process of sensitive space-related mega projects. Perhaps such issues are indirectly addressed by the recommendation he makes with regards to involving "human" arbitrators in the process, in what he calls a "hybrid" form of contracts.[59]

3.2.9 *Governing Digital Property Rights, Living Trees and "Cryptoligopolies"*

Pistor cites Szabo with regards to defining digital property rights which would refer to "a defined space, whether a namespace or physical space that marks the scope of control rights an owner can exercise" and the "first allocation", once coded, is the ultimate and incorruptible proof as to identifying who owns what.[60] Yet, *a beginning is a very delicate time*, and to that regard, Szabo proposes three strategies: (1) a "digital equivalent to the social contract" (i.e. collective decision-making within a clearly bounded community with its own "process governance"); (2) "roots" or "trees" of claims made by the martketplace (i.e. threads of claims amassing the most followers); and (3) "property clubs"[61] whereby self-contained clubs take the role of government (i.e. verifying and enforcing priority rights). In this hypothetical context, antitrust issues could once again be raised with regards to such "property clubs".

A beginning is the time for taking the most delicate care that the balances are correct, however, within DLT governance, the scales seem to be tilting towards heavyweights which, by the way, are growing. Indeed, some technology holders are "more equal than others", which would impede egalitarian distributiveness embedded within the technology's core philosophy. More precisely, the verification processes (labeled as "mining" in relation with cryptocurrency such as Bitcoin, a controversial component of blockchain, is assured by Bitcoin holders themselves and they can earn extra credits the more they mine. However, more mining requires more capacity and therefore, the bigger the holders get, the more Bitcoins they get,[62] although Bitcoins are limited out there since they can be divided.[63] The observation made here is that cryptocurrency has become oligopolistic. Pistor cites De Filippi and Wright: "As of December 2017, only four mining pools controlled over 50% of the Bitcoin network, and two mining pools controlled more than 50

[59] Israel, *supra,* note 2.

[60] Pistor, *supra,* note 25, p. 192.

[61] Pistor, *supra,* note 25, p. 192 citing Szabo, N., "Secure Property Titles with Owner Authority", 1998, publications of the Satoshi Nakamoto Institute, retrieved from: https://nakamotoinstitute.org/secure-property-titles/.

[62] Pistor, *supra,* note 25, p. 200.

[63] Pistor, *supra,* note 25, p. 201.

percent of Ethereum"[64] before concluding that "even the most decentralized of the digital platforms are succumbing to the forces of hierarchy".[65]

3.3 Decentralized Governance Tools: A Toolkit for "Monopolycentricity"?

As stated by Pistor, the decentralized characteristic of disseminated granular ledger technology such as the blockchain is rather paradoxical with respected to its ethos since:

> Like the real world, the digital one too is populated by utopists and realists. In the eyes of the social utopists, one of the greatest attractions of the digital code is that it can be **designed as a decentralized governance system** that will place control over all aspects of life in the hands of individuals. Using digits rather than law to code commitments and social relations is not synonymous with decentralization. To the **contrary**, the scalability of digital **codes allows a few super-coders to establish the rules of the game** for everyone else. Some advances in digital technology, however, have created the possibility of decentralized governance, most prominently among them, blockchain technology".[66] (emphasis added)

Once again, Pistor asserts that such meta-coders could very well end up being corporate "incumbents"[67] down the road, through consolidation and multiple series of mergers and acquisitions (M&As).[68] This raises the question as to which extent really does decentralized governance, in this case, equate with "polycentric" governance which includes decentralization and distributiveness in its nature. Advocating that the ledger fosters opportunities for polycentric governance might unravel a bleak future, in terms of regulation predictability, unless a solid framework built around ethics, applicable jurisdiction, antitrust and transparency is put in place.

3.3.1 Antitrust Issues: An Open Ending

Moreover, the oligopolies mentioned in the previous paragraph are likely to converge into monopolies because both DLT infrastructures and cryptocurrencies are

[64]Pistor, *supra*, note 25, p. 201, citing De Filippi and Wright, "Blockchain and the Law", at Loc. 800 (Kindle Edition).

[65]Pistor, *supra*, note 25, p. 201.

[66]Pistor, *supra*, note 25, p. 185.

[67]Pistor, *supra*, note 25, p. 186.

[68]For example, in the space sector, the strained Planetary Resources start-up agreed to being purchased by blockchain company ConenSys. For more, see: https://consensys.net/blog/news/moonshot-3-0-inside-consensys-space-and-trusat/.

subject to IP claims (e.g., between 70 and 160 patent applications per year[69] and by the end of 2018, the US Patent Office alone had more than a thousand blockchain-related patent applications[70] where:

> Earlier individuals and small firms dominated the filing of patents**; large publicly traded corporations have since taken over the field.** The biggest players in finance are making a major push to legally enclose this **digital gold mine**.[71] (emphasis added)

This “self-contracting” situation signals a propitious environment for further concentration of conglomerates, increasing thus the risk of collusion, concerted practices, abuse of dominance and other anti-competitive behavior as elaborated below:

> In addition, **major incumbent banks have joined forces with tech firms to create consortia that exploit the powers of the digital code,** including **blockchain technology** for the members of these **clubs**. They **use open-source** digital codes, but **don’t necessarily offer open access**. The incumbent “**captains of finance**” have discovered the power of the digital code and are using it to **advance their interest**. Moreover, they are using the legal code to protect the digital work that their hired technologists have crafted for them. How far their advances will go in **enclosing the digital commons** will depend in significant part on the future of **patent and trade secrecy laws**.[72] (emphasis added)

Collusion goes hand in hand with opacity, contrasting thus with ethics pertaining to transparency. When this collusion is about enclosing knowledge itself, the impact escalates exponentially. Therefore, clear transparency measures and policy must be put in place to prevent such risk. In the case of concentrated DLT infrastructure, the solution could unravel through IP regimes leaning towards openness. This could be challenged, though, by new concepts such as “smart IP rights” which could either double or double down the path to transparency.[73]

In Pompidou’s words on space ethics policy, pertaining to IP destined to scientific research, the World Commission on the Ethics of Scientific Knowledge and Technology (COMEST), which is an advisory body of United Nations Educational, Scientific and Cultural Organization (UNESCO), recommends:

---

[69] Pistor, *supra*, note 25, p. 203, citing Chuan, T., “The Rate of Blockchain Patent Applications Has Nearly Doubled in 2017”, July 27, 2017, retrieved from: http://coindesk.com/rate-blockchain-patent-applications-nearly-doubled-2017/.

[70] Pistor, *supra*, note 25, p. 203, citing Chen, M., et al., How Valuable is FinTech Innovation, retrieved from ssrn.com/abstract=3106892 (2018).

[71] Ibid. Enumerated corporations include Goldman Sachs, Mastercard and Barclays.

[72] Pistor, *supra*, note 25, p. 204. DLT infrastructure is prone to concentration through “consortium blockchains”, which, according to DLT advocates, increase their competitiveness through “collaboration”. However, the practices described can easily be qualified as “collusion” in legal terms, from an antitrust perspective. More on such consortium proposals can be found in De Filippi, P., “Blockchain et cryptomonnaies”, Que Sais-Je, 2018, p. 111.

[73] Clark, B., et al., “Blockchain and IP Law: A Match made in Crypto Heaven?”, London, United Kingdom, 2018, retrieved from: https://www.wipo.int/wipo_magazine/en/2018/01/article_0005.html.

> To take all appropriate measures to provide researchers with **free access to scientific data in order to guarantee sharing of knowledge** with a view to **promote scientific progress;** to place scientific outer space data at the disposal of the developing countries; to foster the definition of procedures to **permit sharing of the resulting benefits**, bearing in mind the **legitimate interests** of these countries and acting in the **most equitable** and balanced manner possible.[74] (emphasis added)

3.4 Space Finance

As the commerce of space resources resurfaces iteratively (from the space mining now defunct start-ups[75] to NASA's recent commitment to purchase regolith from commercial entities[76]) space commodities futures exchanges materialize under the form of private entities,[77] numerous questions arise at the level of the applicable law to space cyber resources rights (e.g. temporal use, defined *infra*), traded on a space-based financial platform infrastructure, disseminated through space-based decentralized blockchain technology, protected by IP and all these factors' potential incidence on the emerging *lex mercatoria spatialis*.

The knots and bolts of such theses are fused, for demonstrative purposes, in Israel's "provocative" (in his own terms) scenario below:

> (…) let's imagine **a private, international regime** for **allocating mining rights** in **celestial bodies**. The regime comprises a set of rules, **encoded in a web of smart contracts**, that spacecraft operators may voluntarily contract into. Whether governmental or non-governmental, all operators participate on a level playing field. For the sake of imagination, let's **import the basic bargain of the patent system** for purposes of **allocating mining rights**: an operator that explores a resource deposit receives a twenty-year mining right in exchange for making this information available to the world. To obtain a mining right, let's assume the operator must also lock up $1 million in a **smart contract as a deposit**. If that **operator infringes upon another party's mining interest or otherwise violates the conditions of the contract**, that **value is automatically transferred to the injured party**.[78] (emphasis added)

Undoubtedly, the scenario above touches on unchartered dark waters which need more clarity and transparency in due time, but this confirms that such legal reflections are already on the table of privately led discussions, including space law practitioners, and that they should be met by correspondingly echoing consideration from the part of regulators to sustain smooth co-existence of Space 1.0, 2.0 and 3.0.

---

[74]Pompidou, *supra*, note 38.

[75]Such as Deep Space Industries and Planetary Resources.

[76]"NASA Selects Companies to Collect Lunar Resources for Artemis Demonstrations", Press release, Dec 3, 2020, retrieved from: https://www.nasa.gov/press-release/nasa-selects-companies-to-collect-lunar-resources-for-artemis-demonstrations.

[77]An example of such private initiative is the incorporated Space Commodities Exchange. For more, see: http://www.spacecommoditiesexchange.com/.

[78]Israel (2020), *supra*, note 59. p. 724.

Involving regulators is recommended to ensure transparency and predictability of the law since hybrid contracts involving human arbitrators as oracles means creating law behind closed doors[79] and that there is a risk of the privatization of the law with consequences such as diverging precedents departing, secretly, from *opinio juris*, *stare decisis* or even *jus cogens.*

3.4.1 Commoditizing Resources and IP through Code: Use Case Scenario?

Last, Pistor critiques the "radical market" argument brought forth by Posner and Weyl,[80] according to which, to create a "just society", property rights should be replaced with "contingent use rights". While such "temporal" rights closely equate the "enterprise rights" evoked by Tennen[81] and resonate with the term "use" in the Outer Space Treaty legalese,[82] they could indeed provide for a regulatory framework legitimizing the temporary "safety zones" as proposed by the Artemis Accords and ulteriorly foster a prolific space commerce of digital/cyber space resources through commercializing the rights of using them or the IP attached. Nonetheless, Pistor's caveat addresses the fact that a radical market would exclusively empower price at the detriment of consensus and that the highest bidder takes it all, automatically. The only constant is a "registry" recording the rights and prices for tax purposes. This is reminiscent of smart contracts. Furthermore, besides automatization and registries, another close link might be established between the rights to "use" and tokenization. In the case of non-fungible tokens (NFTs), what is transacted is the right to "use" the rights of the NFT (which in this case are mostly

[79]Pistor, *supra*, note 25, p. 181.

[80]Pistor, *supra*, note 25, p. 230, citing Posner, E. A., and Weyl, G., "Radical Markets", Princeton, NJ and Oxford: Princeton University Press, 2018.

[81]Tennen, L. I., "Entreprise Rights and the Legal Regime for Exploitation of Outer Space Resources,", The University of the Pacific Law Review, V47, I2, 2016., p. 285.

[82]While the term is being used throughout the Outer Space Treaty, for the purposes of this paper, particular reference is directed to articles I and II:

"Article I—The exploration and use of outer space, including the moon and other celestial bodies, shall be carried out for the benefit and in the interests of all countries, irrespective of their degree of economic or scientific development, and shall be the province of all mankind.

Outer space, including the moon and other celestial bodies, shall be free for exploration and use by all States without discrimination of any kind, on a basis of equality and in accordance with international law, and there shall be free access to all areas of celestial bodies.

There shall be freedom of scientific investigation in outer space, including the moon and other celestial bodies, and States shall facilitate and encourage international co-operation in such investigation.

Article II—Outer space, including the moon and other celestial bodies, is not subject to national appropriation by claim of sovereignty, by means of use or occupation, or by any other means."

unique or scarce digital art, but which can enclose other assets in the future). Ownership itself is not the very object of the transaction. For this reason, and according to Scatteia, *supra,* it is more than likely that space resources will become tokenized in the future and that they will be traded both as fungible or as NFTs, depending on the resources systems and units. More precisely, the intangible rights pertaining to such resources are likely to be traded before anything else as the transacted asset will be the rights to "use" the intangible assets pertaining to the said resources. Ownership issues will thus be circumvented since the focus will be placed on "use". This perspective prognosticates that intangible space resources tokenization could lead to a highly speculative market and that regulation will struggle for some time to keep up and contend with. Nonetheless, despite the perception that NFTs offer great benefits besides efficiency, they also come with critical legal problems such as infringements attached to the NFTs, embedded smart contracts with a high residual percentage, and high fluctuation in value. These drawbacks can contribute to extreme levels of market volatility from which it might be impossible for markets to self-correct and act as a lasting destabilizing force on the space economy as a whole.

3.5 Discussion

From all the arguments previously made throughout this chapter, it is expected that space commerce will give to very complex issues and the solutions thereto will to be cutting-edge, at the edge not only of technology, but also of law. Law will be made behind closed doors, via contracting and arbitration, until regulators agree as to new non-binding guidelines and binding legislation. In short, soft law will be likely to govern the evolution of the emerging *lex mercatoria spatialis*, which itself seems to adopt decentralization, comparable to DLT. It follows the elusiveness of both transnational law and code. In this case, Pistor's "code of capital" is ad rem.

Above all, given this nebulous backdrop, regulators to ensure compliance with antitrust and fair competition principles to avoid the earlier mentioned oligopolies, monopolies, and collusion with regards to enclosing knowledge and IP in the space sector and consequently facilitating further anti-competitive behavior through the tokenized space commerce and potentially resulting speculation, which would be detrimental to the sustainability of the space ecosystem. In short, Israel's "private ordering" must not end up in concerted practice that would restrict the application of higher principles of international space law such as freedom of access, equality, and non-discrimination.

Furthermore, according to Pompidou, COMEST believes that:

> (…) every **space policy** must be **based** on the **concept of mutual and reciprocal benefits**, while **safeguarding fair competition** and the **principle of return on investment**. It emphasizes the **importance** of the **role that ethics** must play in the choice of a specific

> project and its long-term assessment from the viewpoint of **human security and economic criteria**.[83] (emphasis added)

Space ethics reserve thus a crucial place for reciprocal benefits, fair competition and return on investment. Acknowledging their importance is key, but efforts as to determining implementation measures have stalled. In terms of space IP rights, scholars urge a call for action:

> A lot was said about the focus of space law on **a commons and equity** and that this is a context, which, when it meets the idea of individual rights as promulgated in intellectual property, really needs to be addressed. The assumption is there will be leakage-I think that was the word I heard-of intellectual property and the stress on individual interests. If space law is to maintain **its focus on equity and a commons approach**, there will need to **be some kind of affirmative action for that to happen**.[84] (emphasis added)

Within the international space community, it is pointed out that efforts are stalling due to "fragmentation". However, on a more positive note, Israel distinguishes fragmentation from multipolarity, which allows for more wiggle room in terms of interpretative approaches, still compliant with outer space law, as follows:

> The **absence of a centralized, authoritative mechanism** for adjudicating divergent interpretations suggests that there will be more **variability in national approaches** than in the Space Law 1.0 paradigm. The **difference between constitutional multipolarity and regime-destroying fragmentation** is in the degree of variability permitted. **Good faith interpretations** may diverge but remain tethered to the Treaty. The **regime bends but does not break**.[85] (emphasis added)

Hence this chapter's recommendation to closely examine the jurisdiction timeline in relation with all the analyzed aspects of space resources commerce, as fleshed out throughout the previous sections, in order to lay out the inherent issues and trends that will influence the unfolding of the future space commerce, while surveying thus the actions to take at each levels of the space sector (Space 1.0, 2.0 and 3.0).

To summarize, the following figure (Fig. 3.2.) answers the questions formulated at the beginning of this chapter as to whether the private sector can escape jurisdiction by adding intangible layers to the commerce of space resources. The answer is that they might indeed do so if contracting down the line, thanks to tokenizing on DLT, is pure code with no legal contract attached. The quickest solution would be to require human oracles in each smart contract (hybrid contracts) and to enforce transparency standards to preclude privatization of the law and anti-competitive behavior such as concerted practice.

---

[83]Pompidou, *supra*, note 38.

[84]Myers (2006), *supra*, note 15, p. 417.

[85]Israel (2020), *supra*, note 59.

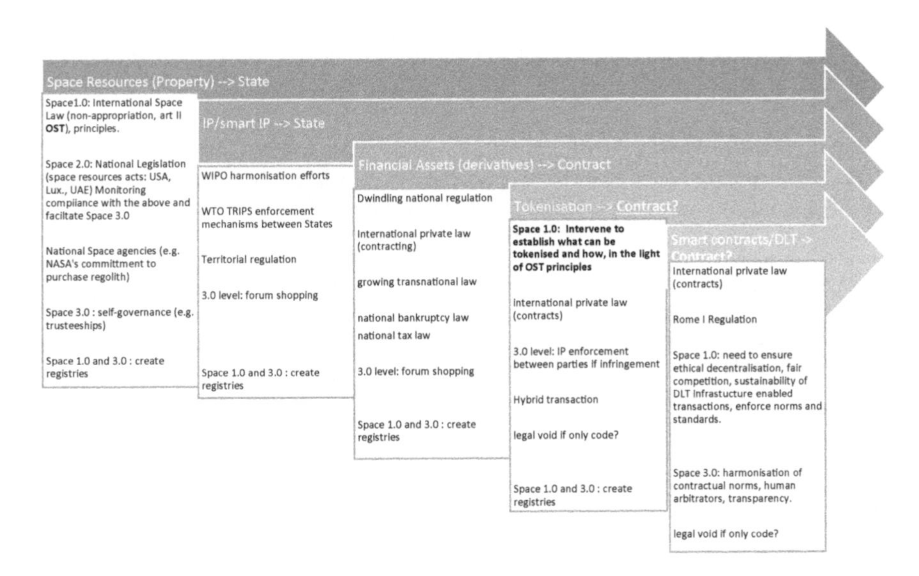

Fig. 3.2. Jurisdiction timeline

3.6 Conclusion

The purpose of this chapter is to determine whether dematerialization of space resources could enable the private sector to escape jurisdiction and if so, how. Thanks to Pistor's line of thinking, as laid out in "Code of Capital", juxtaposed on Israel's Space 3.0 framework, the arguments made throughout this chapter led to answering the key question. The answer is that, in short, if smart contracts of tokenized intangible assets (e.g., IP rights or deregulated financial derivates) pertaining to space resources are pure code and do not include a legal contract, there can be no enforcement. If the smart contract contains a legal contract, though, it is enforceable under international contract law or, in the case of Europe, under Rome I Regulation. Furthermore, to increase chances of enforcement and assignment of jurisdiction, it is recommended to rely on a hybrid contract, involving a human arbitrator as oracle, Last, to prevent privatization of the law, antitrust issues and other unfair competition practices, this chapter recommends transparency measures with regards to the arbitrator's role and contractual standards. Otherwise, tokenization of the space resources market might become very nebulous and blurred by potential resorting to speculative "contractware" on the decentralized DLT platforms.

Acknowledgements The author wishes to thank Professor Dr. Lucien Rapp (University of Toulouse 1 Capitole) for his valuable feedback with regards to this chapter. The author would like to thank Mr. Lucas Mallowan for his crucial editing input.

Maria Lucas-Rhimbassen is a Ph.D. Candidate in Space Law at the Chaire SIRIUS (University of Toulouse, France) where she conducts graduate research on national space legislation, critical space infrastructure resilience, on-orbit insurance and liability, and space antitrust issues. She provides consultation services to public and private stakeholders such as CNES, Airbus and Thales Alenia Space on a variety of topics. Ms. Lucas-Rhimbassen holds graduate degrees in management from HEC Montreal (Canada) and in law from the University of Moncton (Canada) and Le Havre (France) and an executive-level certificate in Strategic Space Law from the IASL at McGill University (Canada).

Chapter 4
Cyber Threats to Space Communications: Space and Cyberspace Policies

Antonio Carlo

Abstract Throughout the last decades, modern society has become increasingly dependent on new technological and digital domains. The strengthening of relations between the space and cyber domains holds the potential to bring disruptive changes. National and international guidelines and recommendations constitute the framework within which the coordinating bodies of both areas advance at national as well as international levels. This article aims to provide an overview of the interconnection between the cyber and the outer space domain from a policy standpoint.

4.1 Interconnections

In 1921, General Giulio Douhet affirmed that air, as the third domain of warfare, would upset the balance of the land and sea domains, undermining the importance of borders between States.[1] Similarly, today, the rise of the space and of cyberspace domains has made borders between States even more intangible.

In the twenty-first century, internet and satellite communications play a vital role in everyday life, considering for instance smartphones and applications that require geographical positioning.[2] Each second, millions of often sensitive data set travel via the internet (or even intranets), allowing for communication through satellite systems, both as agents (satellite internet) and as objects (digital satellite management via intranet systems).

[1]Douhet G. "Il Dominio dell'Aria e altri scritti" (2002) Aeronautica Militare, Ufficio Storico.

[2]Agenzia Spaziale Italiana, "Galileo: Il nuovo programma europeo di navigazione" (2008) Space. Alla scoperta del settore spaziale, supplemento a Il Sole 24 ore, December 2008, p. 3, available at http://doc.mediaplanet.com/projects/papers/Space.pdf.

A. Carlo (✉)
TalTech—Tallinn University of Technology, Tallinn, Estonia
e-mail: ancarl@taltech.ee

A. Froehlich (ed.), *Outer Space and Cyber Space*, Studies in Space Policy 33,
https://doi.org/10.1007/978-3-030-80023-9_4

This means that virtually all critical infrastructure depends on satellite systems. Telecommunications, air and sea transport, financial systems, online banking, military communications and defence systems, scientific monitoring, as well as smart grids[3] are all tied to space infrastructure. The latter includes satellites, ground stations and their interconnections to other terrestrial systems.

Today, nearly 1200 satellites operate in outer space on behalf of 60 states or commercial consortia, whose services are used by users from all around the world. The use of cyberspace is even more extensive, with more than three billion users and an estimated exponential growth.[4]

All of this has resulted in the development of an ever-closer interconnection between two domains, otherwise distinct, namely cyber and space. This continuous linkage is followed by the maturation of an increasing interest in creating legal and political solutions and laws able to regulate and protect these areas which is based on the growing dependence of human activities on these systems, especially with regards to critical infrastructures such as communications. The communication sector alone is in continuous expansion due to its democratisation which has led to the exponential increase of private operators.

Since the launch of Sputnik I, the first Soviet artificial satellite, on 4 October 1957, nations all around the world have gradually begun to increase their presence in outer space to ensure the development of civil, scientific and military applications, and, above all, the implementation of civil and military telecommunications. National and international guidelines and recommendations constitute the framework within which the coordinating bodies of both areas advance at national as well as international level.

Today's satellite capabilities enable the management of increasing portions of civil and military critical infrastructure through IT systems. However, these systems can become the target of cybercriminal actions (whether individual or organised) on a national and transnational level, hence requiring the coordination of law enforcement actions against them. Throughout the past decades, the cyberspace and outer space domains have gained increasing importance in the civil and military environment resulting in them now being recognised as the 4th and 5th domain of warfare respectively. This recognition showcases not only the interest but also the past, present and future investments of private, public and international organisations.

Moreover, both outer space and cyberspace can be considered as "common goods", as recognised in various ways by the international community, making them both domains that cannot be subject to national appropriation.[5] For example, the international treaties on outer space state that the use of outer space «must be

---

[3]Meloni A., Atzori L. "The role of Satellite Communications in the Smart Grid" (2017) In: IEEE Wireless Communications, 2(2), pp. 50–56.

[4]Meyer P. "Outer Space and Cyberspace. A tale of Two Security Realms" (2016) In: Osula A.M., Roigas H. (eds.), International Cyber Norms: Legal, Policy & Industries perspectives, Tallin, NATO CCD-COE, p. 158.

[5]Meyer P. "Outer Space and Cyberspace. A tale of Two Security Realms" (2016) cit., p. 158.

carried out to the advantage and in the interest of all countries […] and must be the province of all mankind».[6] Similarly, the Declaration of Principles adopted by the World Summit on the Information Society (WSIS) in 2005 describes «a people-centred, inclusive and development-oriented information society, where everyone can create, access, use and share information and knowledge».[7]

4.2 Militarisation

The current regulatory framework concerning human activity in outer space dates back to a historical period when the concept and use of outer space substantially differed from today's. This has resulted in an inadequacy to regulate and protect such activities, requiring increasingly urgent modernisation and integration due to the economic issues at stake.

Even before the launch of Sputnik I, the entire international community was concerned about the results of a possible extension of the rivalry between superpowers in outer space. It therefore expressed the idea that space constituted a dimension beyond the sovereignty of states that would not be susceptible to appropriation, a dimension where terrestrial rivalries could not be translated: a *res communis* characterised by a substantial freedom of passage, similarly to that established for the high seas.[8]

The launch of the first satellite, while involving the overflight of the territory of numerous states, did not elicit any protest from any country. Outer space passage therefore appeared free at first, regardless of the purposes of this passage, as long as it was used "for peaceful purposes".

On 13 December 1958, the XIII General Assembly of the United Nations discussed the "peaceful use of outer space issues", during which almost all States used the term "peaceful" as opposed to "military". Emphasising the absolutely innovative nature of outer space activities, the General Assembly stressed the need for international cooperation as long as the exploration and use of space were aimed "exclusively at peaceful purposes".[9] Hence, in 1958, an ad hoc Committee on the Peaceful Uses of Outer Space (COPUOS) was established. This political body was to be composed of two subcommittees, both established in 1961: the Scientific and Technical Subcommittee and the Legal Subcommittee. The purpose of COPUOS

---

[6]Treaty on Principles Governing the Activities of States in the Exploration and Use of Outer Space, Including the Moon and Other Celestial Bodies, entered into force Oct. 10, 1967, Art. I, 18 U.S.T. 2410, 610 U.N.T.S. 205.

[7]Edmon H., Elder L., Perazzini B. "Connecting ICYs to Development" (2014) London, Anthem, pp. 92–95.

[8]Borrini F. "La componente spaziale nella difesa, Soveria Mannelli" (2006) Rubbettino, pp. 21–25.

[9]Boutros-Ghali B. "International Cooperation in Space for Security Enhancement" (1994) In: Space Policy 10 (4), pp. 265–276.

was the development of international cooperation on space matters and the formulation of its regulations, proposing them to the General Assembly for final approval.

Resolution 1472 (XIV) approved on 13 December 1959, introduced the principle that the peaceful use of space and its exploration should be used for the good of humankind and the progress of all States.[10] On the other hand, resolution 1721 A (XVI),[11] adopted unanimously by the General Assembly in 1961, established that outer space and celestial bodies would be open to exploration and use by all States, in accordance with international law and, above all, that these would not be subject to national appropriation.[12]

These resolutions laid out a general legal framework based on programmatic principles that expressed the desire to maintain international peace and security, but deliberately left the normative content undefined. It was believed that this content could be specified later, taking into account political and technological developments. In fact, it would be the historical-political evolution in the relationship between superpowers which, in the absence of a clear and unambiguous definition, determined its current content and interpretation.

Between 1958 and 2008, outer space was the subject of several United Nations General Assembly Resolutions. More than 50 of them were in relation to international cooperation on the peaceful uses of outer space aimed at avoiding an arms race.[13] Four additional treaties on aerospace law were negotiated and drafted by COPUOS: the 1968 Astronaut Rescue Agreement, the 1972 Space Liability Convention, the 1975 Convention on the Registration of Objects Launched into Space, and the 1979 Treaty on the Moon Convention. However, no legal prohibition evolved on the placement of conventional weapons in outer space, resulting in the successful testing of anti-satellite weapons by the United States, the former Soviet Union, and China.[14]

Moreover, one of the most remarkable developments in the field of outer space law has been the adoption of national space legislation. Even countries with very limited outer space activities have developed national policies, considering the

---

[10]G.A. Resolution 1472 (XIV), U.N., 14th Sess., International Co-operation in the Peaceful Uses of Outer Space (12 December 1959). Available at https://www.unoosa.org/pdf/gares/ARES_14_1472E.pdf.

[11]G.A. Resolution 1721 (XVI), U.N., 16th Sess., International Co-operation in the Peaceful Uses of Outer Space (20 December 1961). Available at https://www.unoosa.org/pdf/gares/ARES_16_1721E.pdf.

[12]Janokwitsch P. "The Background and History of Space Law" (2015) In: Von der Dunk F., Tronchetti F. Handbook of Space Law, Cheltenham-Northampton, Elgar, pp. 12–44.

[13]Chesterman S., Malone D.M., Villalpano S., "The Oxford Handbook of United Nations Treaties" (2019) Oxford, Oxford University, pp. 186–194.

[14]Association aéronautique et astronautique de France (3AF) Strategy and International Affairs Commission—Writers' Group "The Militarization and Weaponization of Space: Towards a European Space Deterrent" (2008) In: Space Policy, 24 (2), pp. 61–66.

development and the adoption of national space legislation in the future.[15] Reasons behind this phenomenon include the security of foreign investment, self-positioning as an attractive location for launching space objects and the regulation of domestic space activities.[16]

However, national space legislation tends to differ between States. This is due to the specific needs of each State as well as a range of practical considerations. Such diversity has not been welcomed as it has created confusion and uncertainty about the law applicable to outer space activities. On the one hand, this can lead to inconsistent behaviour by space actors licensed by different national authorities. On the other hand, it may lead to opportunities where private operators apply for a license to conduct outer space activities in those countries that offer the most favourable legislative environment.[17]

However, space technologies can be used for both civil and military purposes. This dual-use aspect has implied a total ban on the use of certain technologies and the impracticality of implementing appropriate measures to verify compliance. This has led to difficulties in reaching arms control agreements. In addition, dual-use technologies make it more difficult to ascertain whether a State is developing a military programme in addition to its civil activities or not, directly affecting the ambiguity of intent surrounding States' actions and thus further stimulating the escalation of political tension.[18]

In practice, the militarisation of outer space involves the placement and development of military weapons and technology in space. As previously stated, the first exploration of space in the mid-twentieth century was at least partly motivated by military ambitions, with the United States and the Soviet Union using this as an opportunity to demonstrate their ballistic missile and other relevant military technologies. Since then, outer space has been used as an operational location for military spacecraft including imaging and communications satellites, as well as intercontinental ballistic missiles that are launched on a sub-orbital flight trajectory. As of 2019, known deployments of weapons stationed in space include only space station armaments and guns such as the TP-82 Cosmonaut survival gun (for post-landing and pre-recovery uses).[19]

At the same time, the exploitation of cyberspace for offensive purposes remains scarce when compared to the overwhelming prevalence of its civil uses.[20] Although initially intended for military purposes, it is the civil use of cyberspace, particularly

---

[15]Esterhazy D. "The Role of the Space Industry in Building Capacity in Emerging Space Nations" (2009) In: Advances in Space Research, 44(9), pp. 1055–1057.

[16]Marbae I. "National Space law" (2015) In: Von der Dunk F., Tronchetti F. Handbook of Space Law, cit., pp. 45–57.

[17]Rosanelli R. "Le attività spaziali nelle politiche di sicurezza e difesa" (2011) Roma, Nuova Cultura—IAI. Available at http://www.iai.it/sites/default/files/iaiq_01.pdf.

[18]Divis D.A. "Military Role Emerges for Galileo" (2002) in GPS World, 13(5), pp. 10–17.

[19]Wong W., Fergusson J., Fergusson J.G. "Military Space Power. A Guide to the Issue" (2010) New York, Praeger.

[20]Meyer P. "Outer Space and Cyberspace. A tale of Two Security Realms" (2016) cit., p. 158.

the internet, that harbours the most criminal activities today. Especially sabotage and espionage have resulted in a new type of conflict namely cyber warfare.[21]

Once again, it is worth reiterating what is expressed in the WSIS Declaration of Principles which highlights the need that «the information society should respect peace», underlining that in a «global culture of cyberspace, security must be promoted, developed and implemented in collaboration with all stakeholders».

In fact, both the use of outer space and cyberspace pose serious challenges to the monitoring and verification of the actions and behaviours of the actors involved. Although a large-scale effort exists to monitor outer space, which is operated mainly by the US military space surveillance network, this is primarily aimed at tracking space debris to avoid collisions. Quite different from this effort is the monitoring of space assets in orbit. It can be argued that in outer space as well as in cyberspace, difficulties exist in verifying compliance with restrictions and in identifying behaviours that violate these restrictions as defined by current agreements.[22]

4.3 Lack of Shared Definitions

As a consequence of many dual-use technologies, the distinction between civil and military purposes is becoming increasingly blurry in both cyber and space.[23] This phenomenon makes it more difficult to define key terminology (particularly war-related terminology) within cyber and outer space, contributing to the lack of or an inadequacy of internationally agreed definitions. This lack also impedes the development of multilateral arms control agreements and discourages cooperation, fostering ambiguity of intent and perceived threats that encourages conflict escalation.

Outer space is a strategic and multidimensional sector and a crossroads for political, strategic, military and economic interests. The picture that emerges, from a legal standpoint, is that of a conventional discipline that has been drawn up in relatively recent times, in which, even if consolidated in the future, grey areas and blurred borders will mostly likely continue to persist.[24]

---

[21]Rajagopalan R.P "Electronic and Cyber Warfare in Outer Space" (2019) Geneve, UNIDIR.

[22]Meyer P. "Outer Space and Cyberspace. A tale of Two Security Realms" (2016) cit., p. 158.

[23]Betza U. "Cybersecurity of NATO's Space-based Strategic Assets" (2019) London, Chatham House. Available at https://www.chathamhouse.org/2019/07/cybersecurity-natos-space-based-strategic-assets.

[24]Brachet G., Deloffre B. "Space for Defence: A European Vision" (2006) In: Space Policy, 22 (2), pp. 92–99; Brachet G. (2004) From Initial Ideas to a European Plan: GMES as an Exemplar of European Space Strategy, in Space Policy, 20 (1), pp. 7–15; Brand S. (2010) Brazil Emerges: A Space Agency With an Eye on Earth, in Tonic Blog. Available at http://www.tonic.com/article/brazil-emerges-a-spaceagency-with-an-eye-on-earth/; Braun F. (2011) Brazil-China Cooperation in Space, in China Digital Times, 10 February. Available at http://chinadigitaltimes.net/2005/01/frank-braun-brazilchina-cooperation-in-space/; Brighel M. (2009) Sicral 1B—le ambizioni spaziali italiane, in Rivista Aeronautica, 85 (3), pp. 84–89.

However, issues such as ambiguity regarding the exact definition of "peaceful purposes" or the absence of a precise delimitation of outer space have led to the adoption of functional solutions. Still, the general nature of the principle of "peaceful purposes" has allowed for the establishment of standards in a high-tech sector characterised by continuous and unstoppable innovation. The lack of stringent limits is therefore not necessarily a negative element but allows for unparalleled flexibility, thus avoiding the risk of rapid obsolescence of definitions. On the other hand, despite a shift in interpretation of some of the principles noted in the outer space treaties, a *status quo* persists due to the general obligation to respect international law and the Charter of the United Nations, particularly concerning the conduct of States in international relations.[25]

From a strategic-military point of view, outer space is proving to be a vital sector for defence and security, as actors are becoming increasingly conscious of the integral part these domains play in military planning and crisis response. As outer space has become increasingly indispensable to the lives of citizens, European institutions have recognised its importance in supporting their policies not only for industrial, economic and political reasons but also for security and defence purposes. The recognition of the significance of cooperation programmes between the European Union (EU) and the European Space Agency (ESA) has even led to a different interpretation of the latter's mandate due to the intrinsically dual-space aspect of space services.

The Lisbon Treaty,[26] which entered into force in 2009, explicitly attributed competence to the EU in space matters albeit together with its member states. It also outlined the vital partnership between the EU and ESA noting, however, their respective independence. From a strictly political-diplomatic and strategic perspective, outer space appears as a centre of gravity for economic, political, military and cultural cooperation between States.[27]

The space dominance of the United States therefore seems to be threatened on the one hand by the expansion of Russian and European space activities, and on the other, by the growth of space activities in emerging countries such as China, India and Japan.[28] The latter are determined to use their political, diplomatic and symbolic potential, but also to acquire technologies capable of accelerating their economic development. Among these, China poses a particular challenge due to a lack of transparency and reliability, especially following the anti-satellite test in 2007. Furthermore, the lack of separation between Beijing's civil and military space activities also brings into question the risks of uncontrolled transfer of technology.

---

[25]UN, Report of the Committee on the Peaceful Uses of Outer Space, New York, UN, 2003.

[26]European Union, Treaty of Lisbon Amending the Treaty on European Union and the Treaty Establishing the European Community, 13 December 2007, 2007/C 306/01. Available at: https://www.refworld.org/docid/476258d32.html.

[27]Bini A "Export Control of Space Items: Preserving Europe's Advantage" (2007) In: Space Policy, 23 (2), pp. 70–72.

[28]Harvey B, Smid H, Pirard T "Emerging Space Powers: The New Space Programs of Asia" (2010) the Middle East and South-America, Praxis.

Establishing export controls of space products and technologies is particularly critical and requires a sacrifice in commercial interest for the benefit of States' national security. However, it is important to establish balanced controls, as evidenced by the case of the US International Traffic in Arms Regulations, which is currently under review, and due to barriers to the transfer of technology between countries participating in cooperative projects.[29]

Ultimately, outer space emerges as an issue of sovereignty and requires a strong involvement of states, assuring autonomy and strategic independence.[30] Space thus arises as a new "stake" in international relations, as it represents an attribute of power and, concurrently, forms an object of negotiation.[31] This is demonstrated by Europe's path towards acquiring independent access to outer space and an autonomous satellite navigation system.[32] Here, too, some questions remain unanswered: For instance, it remains to be clarified how the encrypted positioning signal of Galileo[33] should be used, how the 2004 agreement for compatibility with GPS will be implemented, and how to solve the problem of frequency overlap with the Chinese Beidou system.[34] As far as access to space is concerned, it will be necessary to understand how to face the increasingly aggressive competition in the international launcher market, taking into account the strong government backing of the US industry, and how to ensure the effectiveness of liability management at the international level.

4.4 A Security Affair

The combination of outer space and cyber space policies is based on a simple first-order syllogism:

Cyber systems are subject to attack.
Satellite telecommunications are managed through cyber systems.
Satellite telecommunications are subject to cyber-attacks.

---

[29]Moltz J.C "The Changing Dynamics for the Twenty-First-Century Space Power" (2019) in Journal of Strategic Security, 12(1), pp. 15–43.

[30]Boucher M "Is Canadian Sovereignty at Risk by a Lack of an Indigenous Satellite Launch Capability?" (2011) In Space Ref Canada, 4 January 2011. Available at http://spaceref.ca/national-security/is-canadian-sovereignty-at-risk-by-a-lackof-satellite-launching-capability.html.

[31]Braunschvig D, Garwin R.L, Marwell J.C "Space Diplomacy" (2003) in Foreign Affairs, 82 (4), pp. 156–164.

[32]Bujon de l'Estang F, de Montluc B "Making Space the Key to Security and Defence Capabilities in Europe: What Needs to Be Done" (2006) in Space Policy, 22 (2), pp. 75–78.

[33]Agenzia spaziale italiana "Galileo: il nuovo programma europeo di navigazione" (2008) cit.

[34]Dinerman T "China and Galileo" (2006) Continued, in The Space Review, 21 August. Available at http://www.thespacereview.com/article/685/1; Dinerman T (2009) Galileo and the Chinese: One Thing After Another, in The Space Review, 9 February. Available at http://www.thespacereview.com/article/1307/1.

Therefore, it is important to further explore the use and security of the management information system of all instruments operating in outer space, most importantly satellites. In fact, all technologies related to satellites and other space assets must be regularly updated remotely from Earth. These connections, although protected, could still be attacked and "hacked", giving hackers the ability to access all of the target's systems.[35]

Furthermore, space is changing from a selective environment, managed by wealthy States and the academic world with adequate resources, to one where market forces dominate. Today's technologies bring space capabilities within the reach of nations, international organisations, companies and individuals. Moreover, assets that until a few years ago were owned only by government security agencies, are now in the public domain and are available for purchase on the market.[36]

Cyber-attacks to space infrastructures may include jamming (communications disruption), spoofing (data manipulation), and offensive hacks on communication networks. Other malicious actions could be directed against control systems or mission packages, as well as against ground infrastructure such as satellite control centres.[37] Potential sources of threats further include state offensives, military actions, organised crime seeking large financial returns, terrorist groups seeking to advance their causes, and individuals or groups of hackers seeking personal visibility.[38]

In 2019, ESA launched the project "Funding and support of Space-based services for cyber security", aimed at companies developing innovative products and services in the field of Information and Communications Technology (ITC). This project focuses on satellite-based initiatives to mitigate cybersecurity risks and increase the resilience of existing services, infrastructure and operations. In addition, products that improve end-to-end cybersecurity of space applications are sought. Key project areas are transportation (sea, land, and air, including autonomous vehicles), energy, utilities, critical infrastructure, finance and public safety.[39]

In 2017, the International Group of Experts prepared the Tallinn Manual 2.0,[40] the most relevant non-governmental guide on how existing international law applies

[35]Betza U. "Cybersecurity of NATO's Space-based Strategic Assets" (2019) cit.

[36]Livingstone D., Lewis P. "Space, the Final Frontier for Cybersecurity?" (2016) Chatham House, London. Available at https://www.chathamhouse.org/sites/default/files/publications/research/2016-09-22-space-final-frontier-cybersecurity-livingstone-lewis.pdf.

[37]Carlo A., Veazoglou N. "ASAT Weapones: Enhancing NATO's Operational Capabilities in the Emerging Space Dependent Era" (2019) MESAS 2019, Palermo, Italy. In: Mazal J., Fagiolini A., Vasik P. Modelling and Simulation for Autonomous Systems. 6th International Conference, MESAS 2019, Palermo, Italy.

[38]Rajagopalan R.P. "Electronic and Cyber Warfare in Outer Space" (2019) cit., pp. 6–8.

[39]Duquerroy L. "Cyber Security and Space Based Services" (2019) ESA. Available at https://business.esa.int/sites/default/files/Cybersecurity%20and%20Space%20based%20service_Webinar_Slides.pdf.

[40]Schmitt, M. "Tallinn Manual 2.0 on the International Law Applicable to Cyber Operations" (2017) In Tallinn Manual 2.0 on the International Law Applicable to Cyber Operations. Cambridge: Cambridge University Press.

to cyber activities. The Tallinn Manual established which acts of States violate the general principles of international law in the context of cyberspace. For instance, according to the Tallinn Manual 2.0, a cyber-operation may qualify as "use of force" amounting to an aggression if it entails the necessary scale and effects—a notion used by the International Court of Justice to qualify certain actions as an armed attack. Furthermore, independent projects are currently underway to develop manuals that will further expand on how international laws apply to not only cyberspace but also military space operations.

Although space and cyberspace are two distinct domains, they are closely interlinked and co-dependent: For instance, outer space operations enable a range of operations in the cyberspace just as control segments of systems in outer space require the use of cyber. Ongoing discussions outside the formal multilateral channels are providing ideas and best practices for the implementation of new policies in both domains. While the International Telecommunication Union has affirmed its competence in cyber questions and has developed a reference guide for States to support the development of national cybersecurity strategies, States have not yet come to an agreement on an international regulatory framework for cyber activities. Similarly, wider questions on outer space may also remain unanswered until either States formally agree to a framework of rules or an armed conflict evolves in outer space. This may further indicate that such hostilities will occur in the medium to long-term at the juncture between outer space and cyberspace.

4.5 Policy and Assets Implementation

In 2014, the National Institute of Standards and Technology (NIST) published the Risk Management Framework and the Cybersecurity Framework providing a policy framework for ITC. Even though the NIST cybersecurity maturity standards and guidelines are not directly applicable to the space domain, these are best suited to covering ground-based space infrastructure and assets by assisting organisations in improving their cybersecurity measures and best practices. While efforts have been made to adjust these frameworks to space systems,[41] standards for spacecraft and their associated IoT systems need to be addressed in the near future.

Governance efforts in the space and cyber domains remain highly siloed which has limited meaningful progress. In the past decade, the US has developed various strategy documents covering the improvement of cybersecurity in the space domain, including the 2017 National Security Strategy, 2018 National Cyber Strategy, Space Policy Directive-3, and Space Policy Directive-5 (SPD-5). The latter directive, issued by the Trump administration in September 2020, is the more relevant, representing a government framework that incorporates cybersecurity into

[41]See for instance the 1253F framework by the Committee on National Security Systems Instruction.

all phases of space systems. This directive aims to increase cyber protections for critical US space infrastructure, such as global communications, navigation, and national security applications.

The SPD-5 aims to develop a culture of prevention, active defence, risk management, and the sharing of best practices to establish cybersecurity protocol. This includes «cybersecurity hygiene practices, physical security for automated information systems, and intrusion detection methodologies».[42] Moreover, SPD-5 encourages public and private space operators to share information, best practices and analysis through the Space Information Sharing and Analysis Centre (S-ISAC) operated by the National Cybersecurity Centre (NCC). Even though SPD-5 serves as a high-level policy directive, it should not be understood as a substantive IT governance framework or standard. Therefore, there are urgent issues that need to be addressed immediately, namely the verification and implementation of the assets being adopted in this area and the implementation of the current outlook for the coordination of cybersecurity policies for satellite communication systems.[43]

Both space and cyber activities have their own national and international regulatory frameworks which form the basis for promising future developments, even though these may often be lacking (with respect to the demands that gradually arise) and, above all, are poorly integrated into the international arena.[44] What seems to be absent is an overall vision: A formal coordination between the two areas in terms of policies and assets has not yet been achieved.[45] While analysts and specialists in their cyber and space fields have highlighted the interconnections between outer space and IT activities, pointing to various documents and national policy guidelines, they lack a unified voice and coordination.[46] In fact, both outer space and cyberspace have not yet been the object of strict international regulations aimed at preserving their initially peaceful character.[47] The absence or, at least, the limitation of policy interventions in this regard, constitutes an area of development for all states and supranational institutions, given the widespread recognition of their increasing vulnerability to attacks.[48]

---

[42]Executive Office of the President "Space Policy Directive-5: Cybersecurity Principles for Space Systems" (2020) 09 September. Available at https://www.federalregister.gov/documents/2020/09/10/2020-20150/cybersecurity-principles-for-space-systems.

[43]Ertan A., Floyd K., Pernik P., Steens T. "Cyber Threats and NATO 2030: Horizon Scanning and Analysis" (2020) Tallin, NATO CCD-COE.

[44]ITU "Guide to developing a National Cybersecurity strategy. Strategic Engagement in Cybersecurity" (2018) Geneve, ITU.

[45]European Union Agencies for Cybersecurity "Good Practices in Innovation under National Cyber Security Systems" (2019) Candia, ENISA.

[46]Falco G. "Cybersecurity Principles for Space Systems" (2018) in Journal of Aerospace Information's Systems, 16 (2), pp. 1–10.

[47]Goh Meishan G. "Dispute settlement in international space law" (2007) Leiden-Boston, Nijhoff.

[48]Meyer P. "Outer Space and Cyberspace. A tale of Two Security Realms" (2016) cit., p. 159.

It will have to be observed how the interconnections between elements in cyber and outer space further develop, both in their domain and beyond. Gaining this understanding will be essential to better grasp how these elements interact with each other and how their resilience can be ensured.

Antonio Carlo is currently working at NATO HQ. He is also a PhD candidate at the Tallinn University of Technology specialising in space and cyber, particularly in defence and telecommunication.

Chapter 5
Interdependences Between Space and Cyberspace in a Context of Increasing Militarization and Emerging Weaponization of Outer Space—A French Perspective

Clémence Poirier

Abstract Cyber threats on space systems are becoming an area of interest and concern for space powers and are gradually taken into account in public policies. While most of the literature focuses on Anglo-Saxon countries, it remains to be investigated how cybersecurity risks are considered in French policies. The Space Defence Strategy, released in 2019, is the first French policy document that explicitly recognized the risk of cyberattacks on space systems. This chapter will explore the various interdependences between outer space and cyberspace in French defence policies, France's interpretation of international law in cyberspace and its implications regarding cyberattacks on space systems. Finally, it will look into the French declaratory strategy in case of cyberattacks.

5.1 Introduction

In 2018, French Defence Minister Florence Parly announced that a space defence strategy will be established. She underlined the essential role of space systems for the armed forces as well as the changing context in the space domain: *"It is from outer space that we observe our enemies, their moves, that we find their hideouts and understand their modes of action. It is essential because it is thanks to space that we prepare and plan our operations, fight terrorism, ensure the protection of our deployed forces on all theatres... Today, some States have the means, in space or from Earth, by manoeuvring or by using force, to prevent our access to space or to damage the space capabilities of some countries. Satellites become preys and*

C. Poirier (✉)
European Space Policy Institute (ESPI), Vienna, Austria
e-mail: clemence.poirier@espi.or.at

A. Froehlich (ed.), *Outer Space and Cyber Space*, Studies in Space Policy 33,
https://doi.org/10.1007/978-3-030-80023-9_5

targets".[1] Some of the means mentioned by the Minister could be cyber means. Some countries, not to mention non-state actors, have the capacity to launch cyberattacks against space systems which could disturb military operations, critical infrastructures and compromise governmental communications.

Indeed, in the last few decades, two phenomena have been occurring at the interconnection of space and cyberspace which resulted in the increase of the attack surface, a lower barrier of entry and made satellites new cyber targets. On the one hand, space systems have undergone a digital transformation, going from analogue electronics to digitalised systems. Satellites are increasingly equipped with software defined radios, on-board computers and connected through IP protocols.[2] This trend is rising with inter-satellite links, on-board data processing software, cloud ground stations and other space-based data services. However, it comes with increased risks of cyberattacks and vulnerabilities. On the other hand, outer space has been militarized since the dawn of the space age. This is an old phenomenon that started during the Cold War and with close links with the development of nuclear weapons. It consists in the use of space for military purposes on Earth. Today, a new phenomenon is emerging, the weaponization of outer space, that is to say the placement and deployment of weapons in outer space.[3] In this context, space and cyberspace are interlinked to the extent that space is now militarised and weaponised through cyber means.[4] At the moment, the weaponization of outer space is characterized by discrete threats below the threshold of violence such as cyber or electronic attacks on space systems.[5] This phenomenon is consistent with the militarisation of cyberspace itself in which threats used to come from hacktivists, hackers or criminals looking for financial gains but are now coming from state-actors, their proxies, criminals, terrorist groups, hackers and activists in order to serve their interests.[6]

In order to better understand cyber threats on space systems, it is important to define cyberspace. Cyberspace is often described as composed of three layers: a physical layer (1), a logical layer (2) and a semantic layer (3). Although cyberspace seems only virtual, the first layer is rather tangible. This physical layer, also called a material or infrastructure layer, refers to the equipment, infrastructure and machines such as computers, submarine cables, smartphones and satellites that enable data to

---

[1]Laurent Lagneau, 'Mme Parly: "La Guerre Antisatellite Est Déjà Une Réalité: Il Faudra Compter Avec Nous!"' (*Zone Militaire* 2018) <http://www.opex360.com/2018/09/08/mme-parly-guerre-antisatellite-deja-realite-faudra-compter/> accessed 3 May 2020.

[2]Percy Jackson Blount, 'Satellites Are Just Things In The Internet Of Things' (2019) 42 Air & Space Law.

[3]Xavier Pasco, *Le Nouvel Âge Spatial, De La Guerre Froide Au New Space* (CNRS Editions 2017).

[4]Olivier Becht, Stéphane Trompille, 'Rapport d'information sur le secteur spatial de défense' (Assemblée Nationale 2019).

[5]Ibid.

[6]Percy Jackson Blount (n 3).

flow through cyberspace.[7] It includes servers, USB keys and hardware where the data is stored.[8] This infrastructure can be geographically located and physically destroyed.[9] The logical layer, also called the software layer, consists in the lines of codes in various programming languages and binary information that the machine will transform into readable information for the end user. It also refers to protocols and software such as the TCP/IP protocol that will allow machines to interact with one another and enable the information to disseminate in the form of data packets.[10] The semantic layer, also called the social, cognitive or informational layer, consists in the actual information and data exchanged in cyberspace. It includes end users, their digital identities and interactions.[11]

The French Ministry of the Armed Forces defines cyberspace as *"a global domain composed of the meshed network of information technology infrastructures (including the internet), telecommunication networks, computer networks, processors and integrated control mechanisms. It includes digital information as well as operators of online services".*[12] The French National Cybersecurity Agency (ANSSI), defines cyberspace as *"the communication space that consists in the international interconnection of automated digital data processing equipment".*[13]

In political science, the literature on cyber threats against space systems is very recent. It first tackled the difficulty to understand this new risk at the frontier of two domains, from being described as new types of threats which are difficult to map and define,[14] to being analysed as oversimplified and misunderstood in international relations debates.[15] Authors pointed out the incompatibility of public policies to tackle this threat in a context of increased militarization of outer space and cyberspace[16] or the emerging nature of the threat.[17] Then, the existing literature mostly focused on the lack of consideration for cyber threats on space systems in

---

[7]Kevin Limonier, *Ru.Net, Géopolitique Du Cyberespace Russophone* (Les Carnets de l'Observatoire, L'inventaire 2018).

[8]*'Éléments publics de doctrine de lutte informatique offensive'*, Commandement de la Cyberdéfense (Ministère des Armées, 2018).

[9]Frédérick Douzet, '*La géopolitique pour comprendre le cyberespace*' (Hérodote 2014).

[10]Ibid.

[11]Olivier Kempf, *Alliances et mésalliances dans le cyberespace*, Collection Cyberstratégie (Economica 2014).

[12]'La Cyberdéfense' (*Ministère des Armées* 2020) <https://www.defence.gouv.fr/portail/enjeux2/la-cyberdefence/la-cyberdefence/presentation> accessed 5 May 2020.

[13]*'Stratégie de la France, Défense et Sécurité des Systèmes d'Information'* (ANSSI 2011).

[14]Lorenzo Valeri, *Countering Threats in Space and Cyberspace: A proposed Combined Approach (Chatham House 2013).*

[15]Jason Fritz, *Satellite hacking: A guide for the perplexed*, Vol. 10: Iss. 1, Article 3 (Culture Mandala: The Bulletin of the Centre for East–West Cultural and Economic Studies 2013).

[16]Caroline Baylon, *Challenges at the Intersection of Cyber Security and Space Security* (Chatham House 2014).

[17]Pellegrino Massimo, *Space Security for Europe*, Report n°29 (European Union Institute for Security Studies 2016).

public policies.[18] However, in the French literature, the topic is rarely covered from a policy perspective. Therefore, it seems interesting to study how French policies comprehend and analyse cyber threats on space systems in defence policies, space policies and cyber policies.

5.2 Historical Perspective: Cyber Threats on Space Systems in French Defence Policies

Since the launch of Sputnik in 1957, the militarization and weaponization of outer space have mostly been considered from a kinetic perspective.[19] Space has always been mentioned in France's defence white papers, but the perception of threats and risks have greatly changed over the years. Indeed, during the "first space age", space was mostly thought from a nuclear deterrence perspective since the same technology is used to develop both launchers and ballistic missiles. In 1972, the first Defence White Paper mentioned space, but its role was limited to nuclear deterrence.[20] It stated that only weapons of mass destruction could be used as means of attack in space.[21]

Then, since 1990, the militarization of outer space has been thought from an operational perspective in which satellites became multipliers of power in military operations on Earth (land, sea, air) as it was demonstrated during the Gulf War. The risk of weaponization of outer space was mostly centred on kinetic threats such as Anti-Satellite Weapons rather than cyber threats.[22] In 2008, the French Defence White Paper alerted on the risk of the deployment of weapons in outer space as well as the need of early warning systems to detect ASAT tests and ballistic missiles.[23] The cyber risks on space systems are taken into account but only on the physical layer of cyberspace, that is to say the physical destruction of satellites or ground stations that would prevent access to digital networks and computer tools and thereby disturb military operations. A direct cyberattack on a space system is not mentioned as a potential threat.[24] In 2013, the French Defence White Paper mostly

---

[18]David Livingstone, Patricia Lewis, *Space, The Final Frontier For Cybersecurity?* (Chatham House 2016); Gregory Falco, *Job One For Space Force: Space Asset Cybersecurity* (Harvard Belfer Center for Science and International Affairs 2018).

[19]Beyza Unal, *Cybersecurity of Space-based Weapons Systems* (SGAC 2020). <https://www.youtube.com/watch?v=GDYkd2NzUuo> accessed 14 August 2020.

[20]Béatrice Hainaut, *Stratégie Spatiale en France, De la maitrise au contrôle de l'espace*, 67 (DSI 2019) 44–45.

[21]*Livre blanc* (Ministère des Armées 1972).

[22]Xavier Pasco, *Le Nouvel Âge Spatial, De La Guerre Froide Au New Space* (CNRS Editions 2017).

[23]*Le Livre blanc, Défense et Sécurité Nationale*, Odile Jacob, La documentation française (Ministère de la Défense 2008).

[24]Ibid, 53.

mentioned ASAT weapons and the capacity to detect that threat through Space Situational Awareness capabilities.[25] As a result, the risk of cyberattacks on space systems was mostly overlooked in France's white papers.

It is only recently that the militarization and weaponization of outer space have been perceived from a non-kinetic perspective (cyber and electronic warfare).[26] Without mentioning cyber threats directly, the 2017 National Security and Defence Strategic Review reported on the risk of access denial (DoS/DDoS) on space systems. It also recognized the risk related to the absence of regulation regarding space-based data.[27] In 2018, France publicly condemned the hostile approach of the Russian satellite Luch-Olymp near the French-Italian military satellite Athéna-Fidus in what looked like an attempt to intercept communications. This event brought back the topic of non-kinetic threats on space systems in the French strategic debate. France eventually released its first Space Defence Strategy in 2019 in which cyber threats on space systems are explicitly mentioned.[28]

5.3 The Recognition of Space and Cyberspace as Warfighting Domains

In April 2019, French Defence Minister Florence Parly declared: *"In 2049, war will be permanent and invisible because it is exporting to new types of battlefields that cannot always be perceived by our fellow citizens, whether it is space or cyberspace"*[29] In France, space and cyberspace have both been recognized as warfighting domains just like land, sea and air. This acknowledgement means that the perspective of an open conflict in these domains is possible.

Regarding cyberspace, the 2008 defence white paper recognized cyberspace as the fifth domain of warfare and makes cyberdefence a component of France's defence policy.[30] When the French offensive cyber policy was unveiled in 2019, cyberspace was referred to as *"a new domain"*[31] in Florence Parly's speech and as *"a space of confrontation like any other"* in the policy document.[32]

---

[25]*Livre blanc, Défense et Sécurité Nationale* (Ministère des Armées 2013).

[26]Beyza Unal (n 20).

[27]'Revue Stratégique De Défense Et De Sécurité Nationale' (*Ministère des Armées*, 2017) <https://www.defense.gouv.fr/dgris/politique-de-defense/revue-strategique-2017/revue-strategique> accessed 11 June 2020.

[28]*Stratégie Spatiale de Défense*, Rapport de travail du groupe «Espace» (Ministère des Armées 2019).

[29]'Florence Parly: "En 2049, La Guerre Sera Permanente Et Invisible"' (*L'Obs*, 2019). <https://www.youtube.com/watch?v=BqahmbOySCY> accessed 10 August 2020.

[30]Ministère des Armées (n 24).

[31]Ministère des Armées (n 9) 6.

[32]Florence Parly, *Discours de présentation de la stratégie cyber des Armées* (Ministère des Armées, 2019).

Regarding outer space, the 2017 National Security and Defence Strategic Review recognized outer space as a place of confrontation.[33] As the former head of strategic studies at Thales, Christian Malis, said in 2002: *"Strictly speaking, a space strategy can only exist as soon as outer space will become a theatre of military operations"*.[34] Therefore, the release of the Space Defence Strategy in 2019 marks the acknowledgment of space as an operational domain. Space is about *"operating thanks to space and in space"*, making space a fully-fledged operational domain.[35] Finally, France updated the 2017 Strategic Review by releasing a policy document called "Strategic Update" in 2021. This policy document pairs space and cyberspace as *"acknowledged domains of permanent strategic rivalry, or even conflict"* where some actors are *"taking advantage of the unprecedented accessibility of space as well as the low cost of certain courses of action in cyberspace, often using affiliated non-state intermediaries"*.[36] It outlines that the overall French Defence Strategy has been focusing on these new domains in order to preserve freedom of action. Therefore, the extension of warfare in space and cyberspace creates interdependences which are now recognized in French defence policies.

5.4 France's Interpretation on the Application of International Law in Cyberspace: Implications for Space Systems

First, it should be noted that there is no official and consensual definition of a cyberattack on space systems or cyber weapon in space. Cyber risks on space systems are rarely discussed in international forums. For instance, the UN Governmental Group of Experts (GGE) on Cyberspace and the UN GGE on Space are rarely working together and have not addressed the issue.[37]

Space law does not mention, let alone define, cyberattacks on space systems. Is a cyberattack on a computer on Earth considered as a cyberattack on a space system if it ends up affecting a satellite? Are cyberattacks on the ground segment considered as cyberattacks on space systems? NATO's Tallinn Manual 2.0 is the only document that defines cyberattacks on space systems by distinguishing space-enabled cyber operations and cyber-enabled space operations. On the one hand,

[33]Ministère des Armées (n 28).

[34]Christian Malis, *L'espace extra-atmosphérique, enjeu stratégique et conflictualité de demain* (ISC-CFHM-IHCC 2002).

[35]Florence Parly, *Discours de présentation de la stratégie spatiale de défense* (Ministère des Armées 2019).

[36]'Actualisation Stratégique' (*Ministère des Armées*, 2021) <https://www.defense.gouv.fr/dgris/presentation/evenements/actualisation-strategique-2021> accessed 14 March 2021.

[37]David Fidler, 'Cybersecurity And The New Era Of Space Activities' (*Council on Foreign Relations*, 2018) <https://www.cfr.org/report/cybersecurity-and-new-era-space-activities> accessed 7 May 2020.

space-enabled cyber operations cannot be considered as cyberattacks against space systems if satellites are only used as mean to connect or transfer data instead of relying on terrestrial systems like submarine cables. They do not produce effects or damage in outer space. On the other hand, cyber-enabled space operations can be considered as cyberattacks on space systems. In this case, cyberspace is used to take control, compromise or damage a space object and therefore creates direct effects in outer space.[38] Nevertheless, while France is a member of NATO, the Tallinn Manual is only a research project conducted by NATO's Cooperative Cyber Defence Centre of Excellence (CCDCOE) and does not constitute a policy or law.

Moreover, the application of international law and its principles in cyberspace are the subject of contradictory views at the international level. This is why France released a document outlining the country's vision on the application of international law to cyberspace as well as international law principles. According to this document, France believes that *"a cyberattack could be qualified as an armed attack if it causes substantial human losses, significant physical or economic damage"*.[39] As a result, a cyberattack against a space system could be considered an armed attack as long as it caused significant damage. Also, in this case, article 51 of the UN Charter on self-defence is applicable and allows France to retaliate to an attack.[40]

France also considers that sovereignty applies to cyberspace. There are two approaches: sovereignty as a principle (it is a principle which entails non-intervention, prohibition of the use of force but is not a rule of law) and sovereignty as a rule (sovereignty is considered as a rule of law that should be respected in cyberspace).[41] France's adopted the latter[42] along with the Netherlands, the Czech Republic, Austria and Germany.[43]

Within the sovereignty as a rule approach, there are two doctrines, the first one is that of the penetration of systems and the second one is called the "de minimis" approach. On one side, the "de minimis" approach means that cyber operations have to reach a minimum level of damage to be considered a violation of

[38]Michael Schmitt, Tallinn Manual 2.0 on the International Law Applicable to Cyber Operations, NATO CCDCOE (Cambridge University Press 2017) 269–270.

[39]*'Droit international appliqué aux opérations dans le cyberespace'* (Ministère des Armées 2019) 9.

[40]Ibid.

[41]Przemyslaw Roguski, *Application of International Law to Cyber Operations, A comparative Analysis of States Views* (The Hague Program for Cyber Norms 2020) 4.

[42]Ministère des Armées (n 39).

[43]Przemysław Roguski, 'The Importance Of New Statements On Sovereignty In Cyberspace By Austria, The Czech Republic And United States' (*Just Security*, 2020) <https://www.justsecurity.org/70108/the-importance-of-new-statements-on-sovereignty-in-cyberspace-by-austria-the-czech-republic-and-united-states/> accessed 23 May 2020.

sovereignty.[44] On the other side, the penetration of systems doctrine means that any unauthorised penetration in a computer system or network on its soil consists in a violation of sovereignty.[45] France adopted this doctrine which is detailed in France's document on the application of international law to cyberspace: *"Any unauthorised penetration by a State on French systems or any production of effects on French territory via a digital vector may constitute, at the least, a breach of sovereignty"*.[46]

These various approaches, even within EU or NATO countries, do not encourage a responsible behaviour in cyberspace, undermine the effect of public attributions and could paralyse countries which are victims in case of an attack because they have trouble justifying a retaliation. However, things get even more complicated when it comes to cyberattacks on space systems. While France considers that any attack on systems or network located on its territory is a violation of its sovereignty, satellites are not located in a place where sovereignty is directly exercised, thereby reinforcing the blur around applicable norms in case of cyberattacks against satellites. However, article VIII of the Outer Space Treaty states that *"a State Party to the Treaty on whose registry an object launched into outer space is carried shall retain jurisdiction and control over such object, and over any personnel thereof, while in outer space or on a celestial body"*.[47] Therefore, France's doctrine on the penetration of systems would still apply in case of a cyberattack against one of its satellites. Finally, in a conference on cyber norms organized by the United Nations Institute for Disarmament Research (UNIDIR), the Head of Strategic Affairs and Cybersecurity Department of the French Ministry of Europe and Foreign Affairs declared that France's interpretation of the application of international law in cyberspace would apply in case of a cyberattack on a French satellite.[48]

5.5 Space and Cyberspace as Tools of France's Five Strategic Functions

The 2013 Defence White Paper introduced five strategic functions which were consolidated in the 2017 National Security and Defence Strategic Review: knowledge and anticipation (1), prevention (2), protection (3), deterrence (4) and

---

[44]Przemyslaw Roguski (n 42) 4.

[45]Ibid.

[46]Ministère des Armées 2019 (n 40).

[47]The *Treaty on Principles Governing the Activities of States in the Exploration and Use of Outer Space, including the Moon and Other Celestial Bodies* (UNTS 1967).

[48]Florian Escudié, 'Implementing Cyber Norms, The Challenges of National Coordination' (UNIDIR 2020).

intervention (5).[49] These functions are core principles of France's defence and national security policies. Space and cyberspace are at the service of each principle.

Firstly, knowledge and anticipation enable decision makers to have the necessary information to reduce uncertainty, avoid strategic surprise and allow for strategic autonomy. The function 'knowledge and anticipation' consists of 5 components: intelligence, awareness of operational areas, diplomatic action, foresight, and control of information.[50] The 2013 White Paper particularly underlines the critical role of space systems for this strategic function. Space enables intelligence gathering: *"The first identification of a site of interest or the first perception of a threat is very often obtained through electronic intelligence gathering"*, that is to say radio, telephone or satellite communications.[51] France has a SIGINT satellite constellation named ELISA and will launch three new SIGINT satellites (named CERES) in 2021.[52] In addition: *"In the field of imagery intelligence (IMINT), space capabilities are a priority, since they can identify, precisely locate and target the tangible reality of risks and threats"*.[53] Signal intelligence and imagery intelligence are managed by the Joint Centre for Training and Interpretation of Imagery (CFIII) of the Direction of Military Intelligence (DRM). It analyses space-based data from Earth Observation satellite Helios and Pléiades as well as radars but also images gathered by MALE drones or optical sensors (Reco NG), which are pods attached to Rafale fighter jets that enable to detect and monitor adversaries.[54] It is about creating geo-referenced data, that is to say layering human, signal, and imagery intelligence on one map to monitor adversaries.[55] This centre receives an increasing amount of data. In 2016, DRM estimated that the number of images to process will be multiplied by 10 by 2020. According to General Jean-Daniel Testé, former Joint Space Commander, the workload of imagery analysts will be multiplied by 1000 with the increasing volume of data.[56] This will require a coordination with cyber means, notably artificial intelligence and machine learning as well as data processing software to anticipate adversaries' actions. Moreover, space systems enable target monitoring, detection of activities, evaluation of enemies' actions,

---

[49]Ministère des Armées (n 26) 70.

[50]Ibid 71–74.

[51]Ibid 72.

[52]Laurent Lagneau, 'Le Renseignement Militaire Devra Patienter Pour Sa Future Capacité Spatiale D'écoutes Électromagnétiques' (*Zone Militaire*, 2019) <http://www.opex360.com/2019/10/28/le-renseignement-militaire-devra-patienter-pour-sa-future-capacite-spatiale-decoutes-electromagnetiques/> accessed 4 July 2020.

[53]Ministère des Armées (n 28) 72.

[54]Ministère des Armées (n 28) 75.

[55]François Clemenceau, 'EXCLUSIF. Le Directeur Du Renseignement Militaire Livre Les Secrets Du Géoréférencement' (*Le JDD*, 2020) <https://www.lejdd.fr/Societe/EXCLUSIF-Le-directeur-du-renseignement-militaire-livre-les-secrets-du-georeferencement-791583> accessed 9 July 2020.

[56]Jean-Daniel Testé, 'Compte Rendu N° 48 De La Commission De La Défense Nationale Et Des Forces Armées' (*Assemblée nationale*, 2016) <http://www.assemblee-nationale.fr/14/cr-cdef/15-16/c1516048.asp#P3_69> accessed 6 July 2020.

optimisation and securitisation of itineraries, thereby providing battlefield awareness and a cartography of areas of interest for the armed forces. In the same way, DRM launched the "SYNAPSE" initiative to map the logical and physical layers of cyberspace. It is about establishing a cyber situational awareness by tracing the packets of data in cyberspace in a similar way to how radars would monitor space objects in orbit.[57] It enables to identify and detect cyberattacks more rapidly and attribute them.[58]

Secondly, prevention is the logical consequence of the knowledge and anticipation function to the extent that it consists in acting to avoid the outbreak or worsening of a threat or crisis.[59] Prevention entails the stabilisation of areas whose degraded security could threaten France's national security. It makes the satellite surveillance of areas such as Mali, Libya, Turkey, or Ukraine indispensable. Space systems enable France to have a constant and autonomous capacity to analyse and appreciate events and prevent threats on Earth and in space. Prevention also involves the development of regulations, legal frameworks and confidence building measures through participation in UNCOPUOS and UN Group of Governmental Experts for cyberspace.

Thirdly, protection consists of protecting French soil and its citizens against threats.[60] Space systems enable evacuation plans of French nationals, alerts to the population in case of disaster (detection of earthquake or tsunami).[61] The Armed forces ensure the protection of the territory, coastal areas, air space and outer space. The GRAVES radar ensures the protection of French satellites by detecting threats coming from debris, collisions or other non-cooperative objects.[62] Furthermore, to protect the integrity, availability and confidentiality of military communications, the Joint Direction of Military Infrastructure and Information Systems (DIRISI) is in charge of the cybersecurity of telecommunication satellites.[63]

Fourthly, nuclear deterrence is deemed *"the ultimate guarantee of national security"*.[64] Deterrence is directly linked to space and missile defence to the extent that ballistic missiles go through space before reaching their target. Missile defence consists of intercepting a missile with another one. Today, missiles are hard to

---

[57]Laurent Lagneau, 'La Direction Du Renseignement Militaire Veut Cartographier Le Cyberespace' (*Zone Militaire*, 2019) <http://www.opex360.com/2019/07/05/la-direction-du-renseignement-militaire-veut-cartographier-le-cyberespace/> accessed 6 July 2020.

[58]Ministère des Armées (n 26) 74.

[59]Ibid. 79–80.

[60]Ministère des Armées (n 26) 76.

[61]Thierry Cattanéo *Surveiller l'espace? Enjeux opérationnel et stratégiques pour la défense française,* Conférence La Nuit du Droit (Université Jean Moulin Lyon 3 2019).

[62]'Protéger' (*Ministère des Armées,* 2010) <https://www.defense.gouv.fr/air/defis/fonctions-strategiques/proteger/proteger> accessed 12 June 2020.

[63]'La Direction Interarmées Des Réseaux D'infrastructure Et Des Systèmes D'information De La Défense' (*Ministère des Armées*, 2012) <https://www.defense.gouv.fr/portail/dossiers/l-espace-au-profit-des-operations-militaires/fiches-techniques/dirisi> accessed 15 June 2020.

[64]Ministère des Armées (n 26) 75.

detect because of hypersonic technologies and manoeuvring nuclear warheads. As a result, the best moment and place to intercept a missile is when it is in space because its trajectory is predictable. In addition, deterrence depends on space systems such as early warning systems, EO satellites and radars that enable to observe the enemy. Still, according to Thierry Cattanéo, Head of COSMOS,[65] French nuclear capabilities can be used without space systems, if necessary.[66]

Moreover, the 2013 White Paper and the 2017 Strategic Review only focus on nuclear deterrence as part of this strategic function. Therefore, at first glance, it can be assumed that France does not consider that the theory of deterrence applies to cyberspace. Two theories are opposed: several authors believe that deterrence cannot be applied because the damage caused by a cyberattack are not comparable to those caused by a kinetic attack or comparable to the potential damage of a nuclear weapons. The effects of a cyberattack are not certain and mutual assured destruction is not guaranteed, attribution is more complicated and not borne to State actors, thereby reducing the efficiency of cyber deterrence.[67] Others believe that cyberattacks can cause significant damage and deterrence can be applied to limit the frequency and gravity of cyberattacks. Nuclear deterrence is not linked to cyberspace because French nuclear weapons are not connected and work on a closed and dedicated network. However, a cyberattack on PNT services, ground stations, telemetric data as well as command and control (C2) on which nuclear deterrence rely, could happen. Therefore deterrence, space and cyberspace are increasingly interrelated.

Finally, space and cyberspace are entirely integrated in the fifth core principle: intervention. France depends on space systems for military interventions as 100% of its recent military operations used space systems, particularly the GPS.[68] Space and cyberspace are used for Command and Control, Communication, Computers, Intelligence, Surveillance, Reconnaissance (C4ISR). PNT data enable precision strikes and allows soldiers to coordinate and synchronize their operations, identify targets, and situate themselves in hostile or desertic environments. EO satellites enable the possibility to map conflict areas, provide situational awareness of the battlefield and determine itineraries. Telecommunications satellites enable the management of flows of data from the tactical level to the strategic level, receive orders or transmit drone footage in order to accelerate the OODA loop.[69] Space and

[65]COSMOS (Centre Opérationnel de Surveillance Militaire des Objets Spatiaux) is the Military Space Situation Awareness Center.

[66]Thierry Cattanéo (n 62).

[67]Mariarosaria Taddeo, *The Limits Of Deterrence Theory In Cyberspace*. 33 (Philosophy & Technology 2017) 339–355.

[68]Jean-Daniel Testé (n 56).

[69]OODA: Observe, Orient, Decide and Act.

cyberspace help forces to "acquire and retain operational superiority".[70] As a result, space and cyberspace are interrelated and essential to the application of France's defence policy.

5.6 France's Cyber and Space Policies: From a Defensive to an Offensive Strategy

In July 2019, Defence Minister Florence Parly introduced the Space Defence Strategy. This strategy consists of the identification of current and future stakes and threats in space as well as measures to control this domain and protect French assets.

This strategy takes into account the interdependence between space and cyberspace. Indeed, the strategy recognizes the extension of warfare in both domains. It acknowledges a context in which the combined use of space and cyber technologies by adversaries is reducing the freedom of action of the French armed forces.[71] The cyber threat against space systems is clearly identified. Cyber threats against software components on the ground, space, and user segments are considered as the most likely to occur.

It should be noted that a parliamentary report on the space defence sector which preceded the release of the Space Defence Strategy briefly recognized the cyber threat on space systems and outlined that it was *"identified and taken into account through adapted protection measures"*.[72] However, no further comments were made regarding the type of cyber threats, attack vectors or adversaries. The new strategy clarifies that space systems already integrate protection measures against attacks (including cyberattacks) that could be launched from Earth. It advises that space systems will also have to be protected against attacks that could be launched from space.[73] However, the strategy does not deal with the possibility to launch cyberattacks against other satellites from space.

Moreover, this strategy marks a clear difference with previous policies which were issued by the Ministry of Higher Education and Research. Both the 2012 Space Strategy and the 2016 Space Policy were acknowledging the dependence of the Armed Forces to space but were strictly defensive policies. The Space Defence Strategy is a counter-offensive policy. Defence minister Florence Parly highlighted that fact during the unveiling of the space strategy: *"active defence has nothing to do with an offensive strategy, it is about self-defence. When a hostile action has been detected, characterized and attributed, it is about responding in an adapted*

[70]Ministère des Armées (n 26) 83.

[71]Ministère des Armées (n 29) 24.

[72]Olivier Becht, Stéphane Trompille (n 5).

[73]Ministère des Armées (n 29) 53.

and proportionate way, in compliance with international law principles".[74] In other words, France reserves the right to respond to an attack against its space systems and decides of the moment and the means of the retaliation. As a result, France could use cyber means to respond to an attack against its space systems in order to neutralize the attack itself or neutralize the adversary's operational capabilities (e.g., its satellite TV channel, a connected and/or guided weapon system, etc.).[75]

Furthermore, France considers as essential the capacity to operate in a degraded mode, that is to say without the support of space and cyberspace.[76] The Ministry of the Armed Forces called for the development of alternative technologies or means in case of kinetic or cyber-attacks on space systems.[77]

France's Space Defence Strategy as well as its cyber policies are complementary. French cyber policies also used to be only defensive. It was about protecting critical infrastructures and defend systems in case of an attack with the introduction of the Cyber Defence Pact in 2014 which is centred around 6 priorities: hardening of information systems of the Ministry of the Armed Force (1), supporting research and the cybersecurity industry (2), increasing human resources in the field of cyberdefence (3), building a cyberdefence centre of excellence (4), developing a network of partners with foreign partners (5) and a national community of cyberdefence (6).[78] In terms of capabilities, five years ago, cyber offensive means were very limited. Therefore, the Ministry of Armed Forces had to further develop them before officially adopting a more offensive doctrine.[79] In 2016, COMCYBER, the French Cyber Command, was created and one of its first missions was to contribute to the development of a cyber offensive doctrine. COMCYBER is in charge of the cyberdefence of the computer systems of the Ministry of the Armed Force and is under the authority of the Chief of Staff. It is also in charge of the conception, planification as well as the conduct of cyberdefence operations.[80] In 2018, France published its Cyberdefence Strategic Review, presenting the evolution of cyber threats, policy principles and the hierarchical organization of France's cyberdefence.[81] The LPM 2019–2025 (the defence budget) created a permanent

[74]Florence Parly (n 36).

[75]Ministère des Armées (n 9) 7–8.

[76]Ministère des Armées (n 29) 44.

[77]*Imaginer au-delà, Document d'orientation de l'innovation de défense (Ministère des Armées, 2019) 12.*

[78]'Présentation Du Pacte Défense Cyber' (*Ministères des Armées*, 2014). <https://www.defense.gouv.fr/actualites/articles/presentation-du-pacte-defense-cyber> accessed 23 August 2020.

[79]Laurent Lagneau, 'Environ 40% Des Effectifs Du Commandement De La Cyberdéfense Sont Tournés Vers Les Actions Offensives' (*Zone Militaire*, 2020) <http://www.opex360.com/2020/05/09/environ-40-des-effectifs-du-commandement-de-la-cyberdefense-sont-tournes-vers-les-actions-offensives/> accessed 19 August 2020.

[80]'La Cybersécense' (*Ministère des Armées*, 2018) <https://www.defense.gouv.fr/portail/enjeux2/la-cyberdefense/la-cyberdefense/presentation> accessed 23 August 2020.

[81]*Revue Stratégique de Cyberdéfense* (SGDSN 2018).

posture of cyberdefence (PPC) and allocated additional budget to the cyber domain.[82] In 2019, the Ministry of the Armed Forces released a cyber defensive policy in order to prevent, anticipate, protect, detect, react, and attribute cyberattacks. The same year, it released a cyber offensive doctrine. While only a part of the doctrine is publicly available, it is applicable to all the layers of cyberspace, in which satellites are explicitly mentioned.[83] As a result, space and cyber policies are interrelated and have both progressed towards a counter-offensive posture.

5.7 The Focus on Big Data and Space Situational Awareness in the Space Defence Strategy

The Space Defence Strategy focuses on cyber aspects related to the issue of data processing and the use of artificial intelligence. The processing and analysis of data is essential to detect threats and therefore detect cyberattacks.[84] The issue of Big Data is completely integrated in the statements and recommendations of the Space Defence Strategy. It acknowledges the need to integrate AI to process the increasing volume of space-based data: *"the automatic analysis of space imagery by self-learning algorithms becomes a major challenge for the Ministry of the Armed Forces. To tackle this challenge, AI becomes unavoidable to exploit space data en masse".*[85] Moreover, the strategy plans that those data processing capabilities should be sovereign, developed by the State or trusted operators whose services meet the criteria of availability, confidentiality and integrity required by the Ministry of the Armed Forces.[86] This goal will require substantial efforts to replace foreign data processing software by new French software.

The Armed Forces are dependent on space data and the risks of cognitive overload for analysts are identified in the space strategy. AI is perceived as a solution to mitigate this risk, particularly in the field of Space Situational Awareness (SSA): *"artificial intelligence will play a major role in the processing of data that will be gathered in space: the armed forces will face a 'wall of data'. Their process as well as the automated detection of space objects require massive and united storage capacities, adapted algorithms, and significant computing power".*[87] France bet on AI and Big Data to obtain information superiority on the battlefield and make the most of space-based data. However, France is already relatively late on AI developments compared to other military and space powers.

---

[82] *Projet de loi relatif à la programmation militaire pour les années 2019 à 2025* (Ministère des Armées 2018).

[83] Ministère des Armées (n 9) 2.

[84] Ministère des Armées (n 29) 43.

[85] Ministère des Armées (n 29) 32.

[86] Ministère des Armées (n 29) 48.

[87] Ministère des Armées (n 29) 49.

As a result, the 100 million investments in AI planned in the LPM 2019–2025 (the French defence budget) might not be sufficient.

The incapacity to integrate AI and machine learning in the space domain, particularly in the field of Space Situational Awareness and Space Traffic Management could become a cyber risk in itself. An adversary could target this weakness by sending an overwhelming amount of fake collision alerts or by manoeuvring several satellites in order to paralyse France's detection systems with a high number of alerts. In a context of weaponization of outer space, the incapacity to timely process SSA data would prevent the Armed Forces to have a clear view on the activities conducted in space. It could prevent France from detecting a kinetic attack or a hostile approach in time. SSA and STM relies on the management of metadata and the capacity to transform them into readable and actionable information for military operations.

In addition, the 2018 Cyberdefence Strategic Review plans to exploit Big Data technologies and AI for cybersecurity. AI can be used for simulations, cyber situational awareness, or the detection of attacks and vulnerabilities. The use of AI for cyberdefence purposes is perceived as *"a major stake for France"* in the Cyberdefence Strategic Review. Still, the document also recognizes the lack of research and development in the use of AI for cybersecurity as well as the lack of data to train AI software.[88] It should also be noted that whilst AI can improve cyberdefense capabilities, it is not immune to cyberattacks as it only consists of algorithms and data. A cyberattack could also affect systems (e.g. SSA algorithms) and inject false information, falsify data, or modify the code of the algorithm itself and thereby prevent the Armed Forces from having an accurate SSA.

Moreover, the Space Defence Strategy takes into account the key role of space systems to procure connectivity during military operations: *"Space capabilities will also have to meet the challenge of the armed forces' need for greater connectivity, especially in the air (Rafale F4, drones, FCAS project), on land (the Scorpion programme) and at sea (PANG, the new-generation aircraft carrier), which requires a significant increase in voice communication and data transfer capacities due to the proliferation of connected players and faster data flows"*.[89] To give an order of magnitude, a Rafale fighter jet already produces around 40 To of data per hour and it is expected that future weapon systems such as SCORPION or the Future Combat Air System (FCAS) will generate higher volume of heterogeneous data and require constant connectivity from both space and terrestrial systems. Therefore, the Space Defence Strategy is very ambitious and will probably require additional budget and technological developments to be applied. In comparison, the simple implementation of the Louvois software for the pay of French military personnel already led to technical problems and between 150 and 200 million euros

[88]SGDSN (n 82) 101.

[89]Ministère des Armées (n 29) 49–50.

in additional expenses, that is to say, twice the budget allocated to AI in the LPM 2019–2020.[90]

Overall, cyberspace as a whole is essential to the implementation of France's space defence strategy.

5.8 The Distinctive French Declaratory Cyber Strategy: Acknowledgment and Attribution

The Space Defence Strategy and the cyber doctrine of the French Armed Forces have something in common: they both have been introduced by the public attribution of an attack.

Regarding space, Defence Minister Florence Parly recalled in the introduction of her presentation speech the approach of the Russian Luch-Olympe satellite near the French-Italian satellite Athéna-Fidus which she qualified as an act of espionage.[91] Regarding cyberspace, an Advanced Persistent Threat (APT) attack, attributed to Russia, launched from 2017 to 2018 which targeted the email addresses of the Ministry of Defence's staff as well as the fuel supply chain of the French Navy was highlighted during the presentation of the new doctrine.[92]

The French posture and the declaratory strategy in terms of cyberdefence is quite singular. France almost never communicates on the attacks which target its systems or networks and rarely attributes attacks publicly.[93] In fact, records of cyberattacks on French space systems, space industries and their supply chain are not publicly available. A priori, it could be assumed that space systems are well protected and never targeted. However, attempted attacks or successful attacks are not publicly shared on purpose. According to Kevin Limonier, Head of the Scientific Observatory on Russian-speaking Cyberspace *"the storytelling around an attack is, at the minimum, as important as the attack itself. The lack of data ... makes this topic politically malleable"*.[94] At the policy level, the Strategic Update of 2021 recognizes attribution of cyberattacks as *"a challenge in itself"*.[95]

Indeed, General Tisseyre, head of COMCYBER, believes that public attribution is not an ultimate goal because the accused State will deny these accusations and will ask for proof that cannot be shared publicly without revealing France's

[90]Delphine Mallevoüe, 'Logiciel de Paie Louvois: La Charge de la Cour Des Comptes' (*Le Figaro*, 2014) <https://www.lefigaro.fr/actualite-france/2014/03/06/01016-20140306ARTFIG00350-logiciel-de-paie-louvois-la-charge-de-la-cour-des-comptes.php> accessed 23 August 2020.

[91]Florence Parly (n 36).

[92]Florence Parly (n 33).

[93]Bastien Lachaud, Alexandra Valetta-Ardisson, *Cyberdéfense*, Rapport information n°1141, Commission de la défense nationale et des forces armées, (Assemblée Nationale 2018) 26.

[94]Kevin Limonier (*Twitter 2020*) <https://twitter.com/kevinlimonier/status/1318259989715308545> accessed 26 October 2020.

[95]Ministère des Armées (n 37) 16.

attribution capabilities or without admitting to spying or hacking back to trace the origins of the attack.[96] When attackers are Nation-States, France uses diplomatic channels to confront adversaries.[97] In addition, General Tisseyre believes there is already an excessive number of public attributions and communications regarding cyberattacks conducted by state-actors or their proxies. The consequence of these public attributions is that these groups are pushed to adapt and use means that are more discrete and harder to trace for authorities.[98] As a result, France's declaratory strategy is very different from the "name and shame" strategy usually employed by the Five Eyes (United States, United Kingdom, Canada, Australia, New Zealand).[99]

Moreover, unlike its western allies, adversaries are not explicitly mentioned in the Space Defence Strategy or in the Offensive Cyber Doctrine and Defensive Cyber Doctrine. Potential adversaries are usually mentioned in official addresses, parliamentary auditions or interviews of high-ranking officials. For instance, General Tisseyre declared that COMCYBER was paying close attention to actions conducted by China (particularly industrial espionage), Russia, and Iran.[100]

In the rare occasions when public attributions are made, information is fragmented, and they are made indirectly, that is to say without explicitly stating the name of the country. Additionally, France sometimes takes part in collective public attribution with its EU or NATO allies. For instance, in 2018, the Netherlands, Australia, Canada and the United Kingdom condemned the cyberattack against the OPCW and attributed it to Russia by sharing very detailed information on the people and means used to conduct the attack.[101] France only lent its supports to its allies and joined them in this public attribution but did not make further comments.[102]

Furthermore, while the 2017 Strategic Review only mentions deterrence through the lens of nuclear weapons, the goal of the French declaratory strategy seems to be

---

[96]Laurent Lagneau, 'La Chancelière Allemande Attribue Publiquement La Responsabilité D'une Cyberattaque À La Russie' (*Zone Militaire*, 2020) <http://www.opex360.com/2020/05/13/la-chanceliere-allemande-attribue-publiquement-la-responsabilite-dune-cyberattaque-a-la-russie/> accessed 26 August 2020.

[97]François Delerue, Alix Desforges and Aude Géry, 'A Close Look At France'S New Military Cyber Strategy' (*War on the Rocks*, 2019) <https://warontherocks.com/2019/04/a-close-look-at-frances-new-military-cyber-strategy/> accessed 29 August 2020.

[98]Laurent Lagneau (n 80).

[99]'Le Secrétariat Général À La Défense Clarifie Sa Politique D'attribution Des Cyberattaques' (*Intelligence Online*, 2019) <https://www.intelligenceonline.fr/renseignement-d-etat/2019/05/29/le-secretariat-general-a-la-defence-clarifie-sa-politique-d-attribution-des-cyberattaques,108359078-art> accessed 19 August 2020.

[100]Lagneau, Laurent (n 80).

[101]'La France Accuse (Elle Aussi) La Russie De Cyberattaques' (*La Tribune*, 2018) <https://www.latribune.fr/technos-medias/internet/la-france-accuse-elle-aussi-la-russie-de-cyberattaques-792956.html> accessed 26 August 2020.

[102]'Royaume-Uni—Cyberattaques (04.10.2018)' (*Ministère de l'Europe et des Affaires étrangères*, 2018) <https://www.diplomatie.gouv.fr/fr/dossiers-pays/royaume-uni/evenements/article/royaume-uni-cyberattaques-04-10-2018> accessed 27 August 2020.

deterrence. As strategist Hervé Couteau-Bégarie said, a declaratory strategy is only a way to *"orient the behaviour of the adversary and prevent his error of judgement".*[103] Thus, in her presentation speech of the French cyber strategy, Florence Parly explained that the reason why the offensive doctrine was made public: *"Our potential adversaries have to know what to expect. That is why I decided to make the main principle of our cyber offensive doctrine public, while, of course, keeping private the most sensible elements of it. It is a necessary condition to keep superiority on theatre of operations".*[104] In addition, she affirmed, several times, the capacity of the Armed Forces to attribute attacks.[105]

This dissuasive declaratory strategy is often coupled with the promotion of France's cyberdefence capabilities. General Tisseyre declared *"I would tend to say that France is the strongest Nation in the European Union in terms of cyberdefence, particularly since Brexit".*[106] He added that France's efforts to protect its networks have paid off, as shown by its two victories at the NATO cyber exercise Locked Shields.[107] According to François Delerue, researcher at the Institute of Strategic of the Military School (IRSEM), France is presenting itself as a cyber power capable of deterring attacks by making its cyber offensive doctrine public and promoting its cyberdefence capabilities.[108]

Overall, space and cyberspace constitute the backbone of French military operations. Cyber threats on space systems are taken into account in public policies. France adapted its policy framework to the increasing militarisation of outer space with a counter offensive posture outlined in both the Space Defence Strategy and the Offensive Cyber Doctrine. These policies complement each other to the extent that the former considers cybersecurity risks on space systems and the latter considers satellites as part of cyberspace. Therefore, France could retaliate in case of a cyberattack on its space systems, but it could also respond to any kind of attack through space or cyberspace. Additionally, cyberspace holds a central place in the Space Defence Strategy through the focus on AI and big data for data processing and analysis which directly contribute to France's capacity to assess a situation, decide, and act independently. However, France is relatively late in this field and budgets allocated to these digital technologies is quite limited. The incapacity to

---

[103]Hervé Coutau-Bégarie, *Traité De Stratégie* (7th edn, Bibliothèque stratégique, Economica 2011).

[104]Florence Parly (n 33).

[105]Ibid.

[106]Michel Cabirol, 'Cyberdéfense: "La France Est La Nation La Plus Forte Dans L'Union Européenne".' (*La Tribune*, 2020) <https://www.latribune.fr/entreprises-finance/industrie/aeronautique-defence/cyberdefence-la-france-est-la-nation-la-plus-forte-dans-l-union-europeenne-847379.html> accessed 1 September 2020.

[107]Vincent Lamigeon, '"En Cyberdéfense, Nous Sommes Parmi Les Toutes Meilleures Nations", Selon Le Patron Du Comcyber' (*Challenges*, 2019) <https://www.challenges.fr/entreprise/defence/en-cyberdefence-nous-sommes-parmi-les-toutes-meilleures-nations-selon-le-patron-du-comcyber_685959> accessed 7 July 2020.

[108]François Delerue, Alix Desforges and Aude Géry (n 90).

integrate AI could constitute a cyber risk in itself with regard to data processing. Finally, France opted for a discrete declaratory strategy in which cyberattacks are rarely mentioned or attributed publicly. As a result, no cyber-attack on its space systems have been made public so far.

Clémence Poirier is a Resident Fellow at the European Space Policy Institute (ESPI) in Vienna, Austria. She holds a bachelor's degree in Foreign Applied Languages and a master's degree in International Relations, International Security and Defence from the University Jean Moulin Lyon 3, France. This chapter is a small abstract of her thesis, submitted in 2020, on the interdependences between space and cyberspace in a context of increased militarisation of outer space and the consequences on France's strategic autonomy.

Chapter 6
Cybersecurity and Outer Space: Learning from Connected Challenges

João Falcão Serra

Abstract Humanity is increasingly dependent on space assets, as most critical infrastructure relies on space systems. At the same time, space operations are entirely cyberspace dependent, and space systems present a set of unique challenges that make them especially attractive for hackers. Nevertheless, the national regulatory systems for cyber activities are still underdeveloped, and has a consequence there is still a of lack of proposals for the regulation of cyberspace at the international level. When the time comes to embark on this path, there are negotiation processes in other areas that provided important lessons and that allow not to make unnecessary mistakes. Due to the close relationship between activities in space and cyberspace, which present similar challenges and share others, it is important to analyse and learn from the processes of regulating conventional military activities in outer space. One major lesson is that only a soft law instrument will succeed. Another lesson is that the proponents should not overestimate the aggregating power of a soft law instrument. Lastly, the EU has an advantage to take the lead in the establishment of international norms of behaviour in cybersecurity, as the European bloc inherently deals with this issue on an international level.

6.1 Introduction

Humanity is increasingly dependent on space assets, as most critical infrastructure relies on space systems. To give a few examples, the GPS is used daily from personal navigation to keeping planes from colliding, weather and remote-sensing satellites are important to warn from coming disasters and for the agriculture industry, military early-warning and reconnaissance satellites help States verify compliance with international law and track military manoeuvres.[1] The space

[1]James Clay Moltz, *The politics of space security: strategic restraint and pursuit of national interests* (3rd edn, Standford University Press, 2019) p. 1.

J. Falcão Serra (✉)
Space Law Research Center (SPARC), Nova School of Law, Lisbon, Portugal

A. Froehlich (ed.), *Outer Space and Cyber Space*, Studies in Space Policy 33,
https://doi.org/10.1007/978-3-030-80023-9_6

economy in 2017 was valued at 385 billion dollars and this figure is set to grow.[2] Therefore, governments all around the world recognize the importance of space now more than ever for their citizens, but also for their military capabilities.

There is evidence of significant research and development of a broad range of kinetic and non-kinetic counterspace capabilities in multiple countries, however only non-kinetic capabilities are actively used in current military operations.[3] The US, Russia, China, North Korea, and Iran have all demonstrated to possess cyber capabilities against non-space targets. Although some of these capabilities could be used against space systems, actual evidence of such an attack in the public domain is limited. Additionally, all these countries own electronic warfare capabilities,[4] and not only space is increasingly thought as warfighting domain but there is a focus on asymmetric capabilities and information dominance.[5]

As humanity moved from the analog to the digital age, new challenges which were not previously considered arose. The consequent change of attitudes, starting in the early 2000s departed from a naïve mindset where engineers of space systems thought that their protocols would be too complicated and obscure to crack, the so-called "security through obscurity". And although this change of times does not exclusively affect space systems, they present a set of unique challenges that make them especially attractive for hackers. Gregory Falco defines space systems as "assets that either exist in suborbital or outer space or ground control systems—including launch facilities, for these assets",[6] therefore they include satellites, the ground stations that operate and control them, and the links between them.[7]

---

[2]UNIDIR, 'Supporting Diplomacy: Clearing the Path for Dialogue' (UNIDIR Space Security conference, Geneva, 28–29 of May 2019) <https://swfound.org/events/2019/unidir-space-security-conference-supporting-diplomacy-clearing-the-path-for-dialogue> accessed 21 August 2020 p. 1.

[3]Secure World Foundation, 'Global Counter Space Capabilities: an Open Source Assessment' (2020) p. ix.

[4]The different categories of counterspace capabilities are: *(i)* Direct Ascent: that use ground, air-, or sea-launched missiles with interceptors that are used to kinetically destroy satellites through force of impact, but are not placed into orbit themselves; *(ii)* Co-orbital: weapons that are placed into orbit and then maneuver to approach the target; *(iii)* Directed Energy: weapons that use focused energy, such as laser, particle, or microwave beams to interfere or destroy space systems; *(iv)* Electronic Warfare: weapons that use radiofrequency energy to interfere with or jam the communications to or from satellites; *(v)* Cyber: weapons that use software and network techniques to compromise, control, interfere, or destroy computer systems. See: Ibid, p. xxii and xxiii.

[5]Ibid p. xvii.

[6]Gregory Falco, 'The Vacuum of Space Cybersecurity', (2018, American Institute of Aeronautics and Astronautics) https://www.gregoryfalco.com/publications accessed 31 March 2021 p. 2.

[7]Jana Robinson, 'Governance challenges at the intersection of space and cyber security' (The Space Review, 2016) <https://www.thespacereview.com/article/2923/1> accessed 1 April 2021.

6.2 Challenges

Cyberattacks on space systems can be performed through different means. An IP satellite communications attack is one of them, basically meddling with an IP address to camouflage espionage. Another option is GPS satellite attacks, which can be done through jamming, causing failure on the GPS receiver on Earth to provide a reading, or spoofing, providing data for a fake location. The latter is more dangerous as it appears that the GPS is working as intended.

Additionally, a system that can execute a software-defined spoof attack only costs between 1000 and 2000 USD to build. GPS spoofing is not uncommon, in 2017 there was an incident involving 20 US ships in the Black Sea, and in 2011 the Iranians claimed that they had captured an American drone with this method. Lastly, government satellite and ground space system attacks have also taken place, such as in 2008 when a NASA satellite was under the command and control of the attackers and a security breach in the NASA's Jet Propulsion Laboratory in 2011 which gave full operational control to the hackers.[8]

6.2.1 A Complex Supply Chain Amid a Single Point of Failure

The ITU defines cyberspace as "systems and services connected either directly to or indirectly to the internet, telecommunications and computer networks". As Jana Robinson aptly put: "Space operations are entirely cyberspace dependent",[9] this means that they are subject to its vulnerabilities while also presenting specific challenges.

The complex supply chain of space systems, from the development, to management, use, and ownership of space assets, raises questions on who is liable and responsible for cybersecurity. Additionally, the fact that the lifecycle of space assets is often extensive, means that there are currently aging space systems unable to get patched. That is due to system downtime generally not being an option, another reflection of our great dependency from these systems.

Moreover, space systems represent a single point of failure, that is, if they fail numerous systems will follow the same path. Furthermore, there are various attack vectors from the factories to operators, and supply chains, where one small flaw can cause great damage, as we are dealing with technologies of high precision and volatility. Nevertheless, space asset organisations often lack a specific budget for cybersecurity as most of them do not distinguish between internal IT infrastructure and specialised space systems. NASA employees have been targeted by phishing

---

[8]Gregory Falco, 'The Vacuum of Space Cybersecurity', (2018, American Institute of Aeronautics and Astronautics) <https://www.gregoryfalco.com/publications> accessed 31 March 2021 p. 5–7.
[9]Ibid.

attacks, which if successful can reveal sensitive information. The complex supply chain multiplies this issue, to put it simply, more people equals more risks involved.[10]

6.2.2 Space 4.0: New Actors and Increased Risks

Space 4.0 brings an increased number of private actors to space and a lower entry threshold both in terms of knowledge and financial capabilities. This new era of space is intertwined with Industry 4.0, which is unfolding a fourth industrial revolution of manufacturing and services[11], where everything is digitally connected for instance through the Internet of Things, putting cyber vulnerabilities at even the smallest details of everyday life.

Again, this creates unique challenges for space systems. There is an increase in the use of commercial-of-the-shelf (COTS) technology which augments vulnerabilities. Firstly, it is widely distributed, and a hacker can take hold of the component and analyse it for vulnerabilities. Secondly, they need to be continuously upgraded for security patches, which is often ignored by users. Thirdly, anyone could have contributed to the code behind the open-source technology, meaning possible vulnerabilities or back-doors planted by adversaries.[12]

Gregory Falco argues that CubeSats are especially vulnerable to this because their nature is grounded in COTS technology, making them cheap and more readily available, which is why they are appealing for the new companies of Space 4.0. For start-ups time and funds are usually limited, and therefore there may not be a willingness to spend resources on the development of systems that are robust to cyber-attacks.[13] Not to mention that even in space operations conducted by more traditional space actors, profit margins (although growing) are low, and resources are scarce in the sense that computing power of a satellite might be better employed in something else other than encryption, for instance.[14]

---

[10]Gregory Falco, 'The Vacuum of Space Cybersecurity', (2018, American Institute of Aeronautics and Astronautics) https://www.gregoryfalco.com/publications accessed 31 March 2021 pp. 2–5.

[11]European Space Agency, 'What is space 4.0?' (ESA, 2016) https://www.esa.int/About_Us/Ministerial_Council_2016/What_is_space_4.0 accessed 1 April 2021.

[12]Gregory Falco, 'The Vacuum of Space Cybersecurity', (2018, American Institute of Aeronautics and Astronautics) https://www.gregoryfalco.com/publications accessed 31 March 2021 pp. 4–5.

[13]European Space Policy Institute, 'Brief 26: Cyber Security: High Stakes for the Space Sector' (2018) <https://espi.or.at/news/espi-brief-26-cyber-security-high-stakes-for-the-space-sector> accessed 1 April 2021.

[14]Gregory Falco, 'The Vacuum of Space Cybersecurity', (2018, American Institute of Aeronautics and Astronautics) https://www.gregoryfalco.com/publications accessed 31 March 2021 p. 5.

6.2.3 *Subversion of Integrity*

Although significant resources are being poured into security controls and privacy protections, "cyber-attacks have generally sought to subvert the integrity of political, social, and economic systems, rather than destroy physical infrastructure". This means that usually the hacker's ultimate objective is not a non-reversible non-kinetic attack, such as for example to permanently disable a satellite. Normally, the objective goes beyond the damage in material terms, it intends to sow distrust in the socio-economic fabric of the targeted country. If there is a problem with the timing of the transactions on the financial system, making it stop for two hours, that will create distrust in the financial sector.

This makes sense in the perspective of the attackers. Trust and integrity of systems and institutions are much harder to rebuild than something physical, it is also harder to detect that something is wrong (e.g. theft of information is both harder to be detected and to know the extent the data was compromised, in opposition to the Stuxnet attack designed to stop a centrifuge in the Natanz central in Iran in 2010). Likewise, it is extremely difficult for the defender to know whether the attacker manipulated data or weakened the system integrity. Essential systems and institutions like the global financial system, voting systems, campaigns, and the media are prone to this attacks with potential cascading affects.[15]

6.2.4 *Lack of Regulations*

In the US, industries such as electric systems are regulated by the Federal Energy Regulatory Commission, however cybersecurity standards for space assets are not regulated by any governing body. Furthermore, the regulation of satellites is generally weak, there is no overarching governing body that monitors the specific use of satellites, besides a few standards established in the ITU.[16]

Article 44 of the ITU Constitution establishes that "Radio frequencies and any associated orbits (…) are limited natural resources and that they must be used rationally (…) so that countries or groups of countries may have equitable access to those orbits and frequencies." About the usage of this limited natural resource, Article 45(1) provides that "All stations, whatever their purpose, must be established and operated in such a manner as not to cause harmful interference to the radio services or communications".

---

[15]Neal A. Pollard, Adam Segal, Matthew G. Devost, 'Trust War: Dangerous Trends in Cyber Conflict' (War on the Rocks, 2018) <https://warontherocks.com/2018/01/trust-war-dangerous-trends-cyber-conflict/> accessed 2 April 2021.

[16]Gregory Falco, 'The Vacuum of Space Cybersecurity', (2018, American Institute of Aeronautics and Astronautics) https://www.gregoryfalco.com/publications accessed 31 March 2021 p. 3.

Article 1.169 of the ITU Radio Regulations (RR) clarifies harmful interference as "Interference which endangers the functioning of a radio navigation service or of other safety services or seriously degrades, obstructs, or repeatedly interrupts a radio communication service operating in accordance with Radio Regulations". And Art 1.166 of the ITU RR defines interference as "The effect of unwanted energy due to one or a combination of emissions, radiations, or inductions upon reception in a radio communication system, manifested by any performance degradation, misinterpretation, or loss of information which could be extracted in the absence of such unwanted energy".

The problem is that firstly, there is no distinction between intentional or unintentional harmful interference, although the potential for the latter is generally growing. And secondly, even though there are procedures to deal with harmful interference there is a lack of enforcement measures in case of *intentional* harmful interference.[17]

6.3 International Efforts

There have been efforts to clarify international law's application in cyberspace such as the Tallinn Manual 2.0 on the International Law Applicable to Cyber Operations, however, States continue to conduct cyber operations that violate international law, such as interference with satellite transmissions prohibited by the ITU.[18]

In 2011, Russia and China presented the draft International Code of Conduct for Information Security (Cyber Code of Conduct), updated in 2015. The Sino-Russian proposal provides that States would abstain from using "information and communications technologies, including networks, to carry out hostile activities or acts of aggression, pose threats to international peace and security or proliferate information weapons or related technologies".

Critics have noted that it would justify and permit States to restraint freedom of speech and the free flow of information. Furthermore, they call for language that holds States accountable for cybercriminals acting as agents of the State and for the laws of war to apply to cyberspace, so that hospitals, for instance, are off-limits.

Although the international community seems to agree that there is a need to establish international norms for the cyberspace, the US and British objective is to protect networks and critical infrastructure *while* supporting global efforts to protect

[17]Ingo Baumann, 'GNSS Cybersecurity Threats—An International Law Perspective' (Inside GNSS, 2019) <https://insidegnss.com/gnss-cybersecurity-threats-an-international-law-perspective/ > accessed 2 April 2021.

[18]David P. Fidler, 'Cybersecurity and the New Era of Space Activities' (Council on Foreign Relations, 2018) <https://www.cfr.org/report/cybersecurity-and-new-era-space-activities> accessed 1 April 2021.

the free flow of information, which seem to be the antithesis of what China and Russia proposed.[19]

Therefore, the US, China, and Russia have not agreed on how to approach cybersecurity or address military activities in outer space. Conversely, the negotiations in the UN Committee on Peaceful Uses of Outer Space on guidelines for the long-term sustainability of space activities considered but failed to adopt information-security policies for the terrestrial and orbital parts of space systems.[20]

Satellite systems are vulnerable to both kinetic and cyber-attacks, however the perception and awareness for the risks posed by the first outshines the latter. There have been major efforts and proposals for the regulation of kinetic weapons in outer space which are an answer to the 2007 Chinese direct-ascent anti-satellite test. The counterspace weapons test was largely condemned, raising questions about the sustainability of the space environment and the potential to stimulate a global arms race into outer space. Nevertheless, so far none of the proposals was successful.

6.3.1 Lessons from Past Experiences

Although cyber-attacks and electronic warfare need their own answers, there are valuable lessons to take from the two main proposals, their process of negotiation, and the strategic thinking behind each. The draft Treaty on the Prevention of the Placement of Weapons in Outer Space and of the Threat or Use of Force against Outer Space Objects (PPWT) was submitted in 2008 to the Conference of Disarmament (CD) by Russia and China and updated in 2014. On the other hand, the draft International Code of Conduct for Outer Space Activities (Code of Conduct), is sponsored by the EU and was updated numerous times, the current version being from 2014.

China and Russia advocate for the prohibition on the placement of weapons in outer space through a treaty forged within the designated UN bodies, meaning that its main objective is to curb threats from space systems. This is largely due to the historical threat perceived by such countries, and especially Russia, from the Space Defence Initiative (SDI).[21] On the other hand, the EU focuses on mitigating threats

[19]Timothy Farnsworth, 'China and Russia Submit Cyber Proposal' (Arms Control Association, 2011) https://www.armscontrol.org/act/2011-11/china-russia-submit-cyber-proposal accessed 1 April 2020.

[20]David P. Fidler, 'Cybersecurity and the New Era of Space Activities' (Council on Foreign Relations, 2018) < https://www.cfr.org/report/cybersecurity-and-new-era-space-activities> accessed 1 April 2021.

[21]A program that envisaged an orbital defence arrangement capable of intercepting nuclear weapons, envisaging "the deployment of thousands of orbital weapons". Cit. James Clay Moltz, *The politics of space security: strategic restraint and pursuit of national interests* (3rd edn, Standford University Press, 2019) p. 176.

to space objects and argues for voluntary transparency and confidence-building measures.[22]

6.3.1.1 Hard Law Versus Soft Law

One major difference between the two approaches is the first focuses on hard law and the second on soft law. To operationalize an arms control regime, precise terms and effective means of verification are essential because the rationale behind such agreements must be based on a win–win situation, whereby the national security of the States involved is enhanced by such regime.

The PPWT failed to do so. Not only it lacks verification mechanisms but it defines space weapon as "any device placed in outer space, based on any physical principle, which has been specially produced or converted to destroy, damage or disrupt the normal functioning of objects in outer space, on the Earth or in the Earth's atmosphere, or to eliminate a population or components of the biosphere which are important to human existence or inflict damage on them".[23] Furthermore, Article I (d) provides that "A weapon shall be considered to have been 'placed' in outer space if it orbits the Earth at least once, or follows a section of such an orbit before leaving this orbit, or is permanently located somewhere in outer space".

This definition is ambiguous. By mentioning that a space weapon must be placed in outer space, it automatically leaves out both ballistic missiles[24] and any Earth-based weapon, which China and Russia justified due to the difficulty of verifying their development and effective location.

Furthermore, there is the issue of dual use of space objects, meaning that they can be used both for military and civilian purposes. For example, a GPS satellite can be used to provide situational awareness for a civilian in a city or for the military on the battlefield. In a legal perspective, this makes it hard to establish appropriate means of verification and regulation. Unsurprisingly, dual-use space objects have been omitted from the abovementioned definition since they have not

---

[22]Benjamin Silverstein, Daniel Porras, and John Borrie, "Alternative Approaches and Indicators for the Prevention of an Arms Race in Outer Space" (UNIDIR Space Dossier 5, UNIDIR, 2020) < https://unidir.org/publication/alternative-approaches-and-indicators-prevention-arms-race-outer-space> pp. 9–11.

[23]CD/1985, 'draft treaty prevention of the placement of weapons in outer space and of the threat or use of force against outer space objects introduced by the Russian Federation and China' (12 June 2014), Article I (c).

[24]Which can be used to target satellites and although they go through space they are not *placed* in space. See: Jinyuan Su, The "peaceful purposes" principle in outer space and the Russia–China PPWT Proposal, Space Policy, Volume 26, Issue 2, 2010, Pages 81–90, ISSN 0265–9646, https://doi.org/10.1016/j.spacepol.2010.02.008, p. 85.

been "specially produced or converted" to damage other space objects.[25] This has strategic implications, for example, the US believes that China relies on the ambiguity of dual-use purposes as part of the broader "Military-Civil Fusion Development Strategy".[26]

This issue is also present in the cyberspace. A country's critical infrastructure can be compromised solely by using commercial software, making it impossible to devise means of verification for the technology itself. Therefore, the dual-use nature of both cyber and space technologies dilutes the line between non-military and military usage.

Furthermore, the characteristics of cyber-attacks which may not rely on jamming or spoofing but solely in manipulating information of the network-accessible systems, makes it so that any component of an integrated system may fall victim of such. This type of attacks can be conducted from everywhere, with no special hardware and the perpetrators can easily hide their identity, further making it difficult to create an adequate system that provides transparency.

Therefore, it is virtually impossible to employ traditional arms control approaches that rely on verification and transparency mechanisms that cannot be achieved. In this sense, Transparency and Confidence-building Measures (TCBMs) can still be a venue of mitigation on how to address the use of these technologies.[27] This was largely the thinking behind the Code of Conduct.

The Code of Conduct was negotiated outside the CD and therefore its nature is different from the PPWT in the sense that it is not a pure arms control treaty, having an important environmental component to it, thus giving priority to kinetic ASAT weapons that generate debris. It is in line with the EU position on establishing rules of behaviour (rather than evaluating the legality of weapons in outer space) and

---

[25]On the dual-use dilemma: "One delegation holds that it is not easy to identify what is or is not a weapon in outer space. The logic is that anything in outer space with the ability to alter its trajectory, including any of the current meteorological, communications, remote-sensing, or navigation satellites currently in orbit, could be a weapon and any of these could, in principle, have its orbit altered so as to collide with another satellite, with obviously harmful results to the target. The same delegation argues that the inability to define space weapons is the main barrier to a treaty" Cit. CD/1818, para 39. Also, as Frans von der Dunk metaphorically and eloquently explained: "On my way to boarding the plane (…) I was not allowed to take, inter alia, any knife with me into the aircraft. Mind you: not just knives specially produced or converted to wound or kill people, but any knife, including knives specially produced or converted for example to cut bread—because, obviously, also those could wound or kill people on an aircraft, and there would be little upfront guarantee about absence of malicious intent (…) to use any knife for such purposes." Frans von der Dunk, Cutting the bread, Space Policy, Volume 29, Issue 4, 2013, Pages 231–233,ISSN 0265–9646, https://doi.org/10.1016/j.spacepol.2013.10.001.

[26]Office of the Secretary of State, 'Military and Security Developments Involving the People's Republic of China 2020',(Annual report to Congress, 2020) <https://media.defense.gov/2020/Sep/01/2002488689/-1/-1/1/2020-DOD-CHINA-MILITARY-POWER-REPORT-FINAL.PDF> pp. 18–23.

[27]Jana Robinson, 'Governance challenges at the intersection of space and cyber security' (The Space Review, 2016) <https://www.thespacereview.com/article/2923/1> accessed 1 April 2021.

avoiding conflict in outer space.[28] Transparency is defined as "the degree of openness in conveying information and a device of strategic negotiations signalling the trustworthiness of the actor in negotiations".[29]

The Code of Conduct failed for two main reasons. Firstly, the drafting process was done outside the COPUOS and the CD, fearing that those UN-sanctioned multilateral bodies could stall negotiations.[30] However, this approach alienated the BRICS and NAM countries which argued the process was not inclusive enough. Secondly, it hinged too much on the soft part of "soft law". One of the great advantages of soft law instruments is the flexibility not only from the fact that it is not legally binding, giving leeway to negotiations, but the subproduct of that being that the terms and definitions will not carry the same weight. In other words, controversial issues that would normally kill a legally binding treaty can be brushed over in a soft law instrument.

However, the Code of Conduct took this too far. The scope of the Code is overstretched and too ambitious, it deals with matters of national security that are deeply important for the States (where even a soft law instrument carries considerable weight) with an environmental lens. This created an existential crisis.

For instance, Article 6.1 provides that States resolve to "share their space strategies and policies, including those which are security-related", as well as their major outer space research and programmes, and may consider sharing information on data relevant to governmental and non-governmental entities of other subscribing States, particularly on phenomena that may pose a hazard to spacecraft.[31] This would be seen as a normal transparency and confidence-building measure. Yet, the negotiating parties recognised the insecurity that sharing sensitive information with non-participating States would bring. Consequently, the Code implements restrictive information sharing with "other subscribing States" instead of open sharing.

---

[28]Stacey Henderson, 'Arms Control and Space Security'in Kai-Uwe Schrogl (eds), *Handbook of Space Security* (Springer 2020) pp. 12–13. And also, Delegation of the European Union to the United Nations, 'EU Explanation of Vote—United Nations 1st Committee: No First Placement of Weapons in Outer Space' (2017, New York) <https://eeas.europa.eu/delegations/un-newyork_en/53334/EU%20Explanation%20of%20Vote%20%E2%80%93%20United%20Nations%201st%20Committee:%20No%20First%20Placement%20of%20Weapons%20in%20Outer%20Space>: "(…) the EU and its Member States believe it would be more useful to address the behaviour in, and use of, outer space in order to advance meaningful discussions and initiatives on how to prevent space from becoming an arena for conflict and to ensure the long-term sustainability of the space environment".

[29]C. Ball, 'What is transparency?' (Public Integrity 2009) *apud* Jana Robinson, Transparency and confidence-building measures for space security, Space Policy, Volume 37, Part 3, 2016, Pages 134–144, ISSN 0265–9646, https://doi.org/10.1016/j.spacepol.2016.11.003 p. 134.

[30]Michael Krepon, 'Space Code of Conduct Mugged in New York' (Arms Control Wonk, 2015) <https://www.armscontrolwonk.com/archive/404712/space-code-of-conduct-mugged-in-new-york/> accessed 1 November 2020.

[31]Council of the European Union, 'Draft International Code of Conduct for Outer Space Activities', (31 March 2014), Article 6.2 (Code of Conduct) <https://eeas.europa.eu/archives/docs/non-proliferation-and-disarmament/pdf/space_code_conduct_draft_vers_31-march-2014_en.pdf>.

This goes against the spirit of the Code on "enhancing the safety, security, and sustainability of outer space activities" [32]considering that dangerous activities in outer space potentially affect all space-faring States, even the ones outside the Code, and that therefore "space law generally seeks to maintain a focus on the free and open sharing of information for the benefit and safety of all nations". Furthermore, the fact that States can provide varying amounts of data is also a major setback for the Code since the information provided can be insufficient, inaccurate, and irregular.

Consequently, although a soft law instrument, the fact that it deals with matters of security makes it carry significant weight, upping the demands for verification, clarity of concepts, and transparency that would take it closer to an hard law instrument, making States very reluctant to accept it, especially when the drafting process was not conducted in openness.

The same mistakes should be avoided by any international soft law agreement on cyberspace and space operations that deal with matters of security. One must consider the scope of the proposal in proportion to the actors it wants to engage, in this sense the European Union has an advantage which will be addressed below.

6.3.1.2 Strategy and Deterrence

This brings us to this next sub-chapter. Not everything can be explained strictly through legal analysis, the proposal might be perfectly fine legally speaking but not be strategically viable for the State. In international relations neorealists argue that the world is anarchic, and States can only trust on their own ability to survive (self-help). Therefore, the best way to do so is by maximizing their security. In this sense the amount of power, especially in relation to other States, is crucial.

As such, States are seen as units whose capabilities are measured by military and economic factors.[33] Considering that the US is "unequivocally ahead" in terms of space technology, in light of the neorealist school of thought, the fact that China and Russia proposed the PPWT makes sense given that it proposes banning the deployment of sophisticated space-based weapons, while ignoring the development of less sophisticated direct-ascent ASAT technology. For the US this would certainly mean a relative loss of power, and thus it also makes sense that it rejected it.[34]

On the other hand, neoinstitutionalism fits well in the above sub-chapter on hard law versus soft law. It recognizes the anarchical structure of the international system but argues for a system of interdependence between States in order to cope with it by maximizing their utility through mutual interests. It is the antithesis of the zero-sum game logic whereby a win for State A is necessarily a loss for State B.

---

[32]Cit. Ibid, Preamble.

[33]Max M. Mutschler, 'Security Cooperation in Space and International Relations Theory' in Kai-Uwe Schrogl et al. (eds), *Handbook of Space Security* (Springer 2015), pp. 42–54, p. 48.

[34]Ibid, p. 49.

One of the great problems, however, is the fear of cheating, especially when dealing with security. The main solution for this issue is the creation of a regime that sets up rules that define cheating and help verify compliance, as "no State wants to abandon the development of space weapons only to find out that other States have developed these technologies".

The natural high degree of interdependence provided by the space environment —where one action affects all—provides a good basis for the development of cooperation mechanisms for the implementation of such a regime.[35] Yet, space also provides particularly hard obstacles for effective cooperation. Firstly, the space environment makes it difficult to provide empirical proof of compliance. Secondly, the different interpretations of what a space weapon is, plus the dual-use conundrum, undermines a common understanding between the parties on what would constitute a breach of the arms control regime. Therefore, it is virtually impossible to establish a cooperation regime on arms control.

As the main space faring nation, historically the US relies on deterrence to protect itself, hence it has been an integral part of US policy to oppose anything that impairs its ability to maintain that status.[36] Therefore, from the US point of view, anything that diminishes its current and future capabilities (such as an arms control regime) must provide as much security as the *status quo*. The matter of fact is that the failed adoption of the PPWT and the Code of Conduct seem to have confirmed the US doubts about changing the current strategic and legal framework.

The proposals put forward would not enhance the security of the US enough to replace the *status quo*. Likewise, these proposals would need the US adoption to make them viable, as both wanted to achieve a high degree of international acceptance. Therefore, the US reluctance to join the efforts for the adoption of a Code of Conduct and its rejection of the PPWT proved to be a sobering moment for a whole range of other States, not just the US.

Accordingly, today, the main strategy of the US to preserve its security in outer space is still deterrence.[37] Space deterrence can be defined as "deterring from the attack on space objects and all means supporting space activities, undertaken to interrupt their operations temporarily or permanently".[38]

---

[35]Ibid, pp. 50–51.

[36] Stacey Henderson, 'Arms Control and Space Security' in Kai-Uwe Schrogl (eds), Handbook of Space Security (Springer 2020), p. 14, "The US repeatedly stated that it will oppose any new legal regime or other restriction that seek to prohibit or limit US access to or use of space and arms control agreements cannot impair the rights of the US to conduct activities for US national interests.".

[37]Christopher Ford, 'Whither Arms Control in Outer Space? Space Threats, Space Hypocrisy, and the Hope of Space Norms' (Center for Strategic and International Studies, Washington, DC, 6 April 2020) <https://2017-2021.state.gov/whither-arms-control-in-outer-space-space-threats-space-hypocrisy-and-the-hope-of-space-norms/index.html> accessed 1 November 2020.

[38]Cit. Rafał Kopeć, Space Deterrence: In Search of a "Magical Formula", Space Policy, Volume 47, 2019, Pages 121–129, ISSN 0265–9646, https://doi.org/10.1016/j.spacepol.2018.10.003, p. 123.

Nevertheless, the US also knows that the strategy of deterrence cannot by itself guarantee its security, given that the outer space environment is based on interdependence and thus no country can guarantee space sustainability alone.[39] The US, recognizing that none of the previous efforts are managing to curb the emerging security problems, is working to develop approaches to outer space norms that will help improve predictability and collective "best practices" alongside deterrence.

It seeks to develop verifiable norms of responsible behaviour, to avoid situations of heightened tensions. Meaning that they are ought to be specific enough to check for compliance, one recent example of what would be considered an irresponsible behaviour given by the US, concerns proximity operations with satellites of other space faring nations without prior consultations.

The strategic realities are as important as legal considerations, they are both strictly related. Any international effort that seeks to approach cybersecurity must take this into account, it does not matter if a proposed document is of sound reasoning when the parties do not hear the same. Better then to limit the expectations, the scope and the nations involved, carefully evaluate the strategic thinking behind the States, and tentatively build on practices one by one, instead of presenting grand solutions and a one size fits all.

6.4 National Efforts

So far, "neither international law nor diplomacy has grappled effectively with space cybersecurity",[40] the reason may be due to the fact that the national regulations and principles of the States are still underdeveloped. Before doing so, States do not have a solid base from which to venture into the international arena. There have been instances of space assets being compromised by cyberattacks, which demonstrated that even well-funded space projects lack the appropriate cybersecurity to defend against space hackers.

6.4.1 Cybersecurity in the USA

As military doctrine develops the concept of multidomain operations (MDO), which is understood as joint operations conducted across multiple domains, cyber space operations become a "crucial and inescapable component of military space

[39]Max M. Mutschler, 'Security Cooperation in Space and International Relations Theory' in Kai-Uwe Schrogl et al. (eds), *Handbook of Space Security* (Springer 2015), pp. 42–54, p. 50.

[40]David P. Fidler, 'Cybersecurity and the New Era of Space Activities' (Council on Foreign Relations, 2018) <https://www.cfr.org/report/cybersecurity-and-new-era-space-activities> accessed 1 April 2021.

operations and represent the primary linkage to the other warfighting domains".[41] As such, 80% of US military operations rely on commercial satellites for non-critical missions.[42]

Since 2015, there is a greater drive for the call for reform within NASA, this is not only relevant due to importance of the organisation but also its influence on smaller organisations that follow its standards and best practices. This has been achieved through better access control management, more specialised security workforce (creating the Cyber Defence Engineering and Research Group to specifically address missions systems such as Mars Science Lab), fostering a security culture, beginning to encrypt data while stored and during transfer, and developing custom security tools to help system architects and developers better conceptualise vulnerabilities. However, efforts from private space companies like SpaceX and Blue Origin are kept secret.[43]

Although there are significant efforts, so far, in the US, there is little guidance to develop cybersecurity, one of the few examples of such being the Cybersecurity Information Sharing Act signed into law by President Obama. One recent development is the Space Policy Directive 5 (SPD-5), which establishes important principles in the pursuit of developing national efforts to regulate and create norms of behaviour for the cyberspace.

Yet, it is non-binding, and it does so generally speaking, with no mention of specific US systems. The reason for this limitation is within one of the principles it establishes: that the adoption of the proposed cybersecurity measures should be without undue burden on the mission requirements. This correlates with the notion of Space 4.0 and the low entry barriers. In effect, what is being told here is that these barriers should remain low enough so not to amper innovation.

6.4.2 Cybersecurity at the EU Level

The 2014 EU Cyber Defence Policy Framework aims to protect communications which involves both satellite and terrestrial, in the context of the Common Security and Defence Policy actions. Additionally, the 2017 Diplomatic Toolbox, which enables sanctions against individuals responsible for cyberattacks, could be implemented in case of a cyberattack against a European satellite.

---

[41]United States Space Force, 'SpacePower Doctrine for Space Forces' (2020, Space Capstone Publication) <https://www.spaceforce.mil/Portals/1/Space%20Capstone%20Publication_10%20Aug%202020.pdf> p. 7.

[42]European Space Policy Institute, 'Brief 44: SPD-5: towards a coordinated approach on space cybersecurity' (2021) <https://espi.or.at/news/espi-brief-44-spd-5-towards-a-coordinated-approach-on-space-cybersecurity> accessed 1 April 2021.

[43]Gregory Falco, 'The Vacuum of Space Cybersecurity', (2018, American Institute of Aeronautics and Astronautics) https://www.gregoryfalco.com/publications accessed 31 March 2021 p. 7–8.

But there is no EU legislation or policy dedicated to the cybersecurity of space systems and even information-sharing is still limited. And although the 2020 EU Security Union Strategy recognises space systems as essential services that should be protected against cyber threats, the EU Agency for Cybersecurity has never mentioned space systems in its analyses as of September 2020.

On the national level, at least in comparison with the USA, the scenario is not much better. Since 2019 a number of countries recognised cyber threats on space systems, but as we have seen above, in the US the debate as moved beyond that. And this can start to affect the competitiveness of EU companies in relation to the American ones. With the increasing number of cyber-attacks, cyber security is now a strong component of non-price competitiveness.

Right now, the EU and its Member States are dependent on vulnerable space systems. Despite the interdependence on space programs at the EU level, the EU itself is limited to act on security matters which remain mostly addressed at the national level.[44]

6.4.3 Recommendations

Academics and organisations have outlined important recommendations on the regulation of cyberspace. Policymakers should be proactive, not reactive, especially considering the volatility of the space environment and to avoid overregulating after as a result of an incident. The concept of critical infrastructure should include underlying systems meaning that space systems should be considered as such, so to increase security standards. Additionally, it is important to develop a security culture in all of the space ecosystem.

Moreover, there should be assigned responsibility and liability for cybersecurity, encouraging responsible parties to take necessary measures to secure their systems. Space asset organisations should also be accountable for cybersecurity, the State can do it by requiring that the contracting space organisation for a given project as to comply with key performance parameters pertaining to cybersecurity.

On that note, the obligation of reporting all cyber incidents that have impacted or could impact national security should be expanded to space asset organisations. As Gregory Falco proposes, a Space System Information Sharing and Analysis Centre should be established to share threat information across space system agencies and space asset organisations.[45]

[44]European Space Policy Institute, 'Brief 44: SPD-5: towards a coordinated approach on space cybersecurity' (2021) <https://espi.or.at/news/espi-brief-44-spd-5-towards-a-coordinated-approach-on-space-cybersecurity> accessed 1 April 2021.

[45]Gregory Falco, 'The Vacuum of Space Cybersecurity', (2018, American Institute of Aeronautics and Astronautics) https://www.gregoryfalco.com/publications accessed 31 March 2021 pp. 9–11 and European Space Policy Institute, 'Brief 26: Cyber Security: High Stakes for the Space Sector' (2018) <https://espi.or.at/news/espi-brief-26-cyber-security-high-stakes-for-the-space-sector> accessed 1 April 2021.

Specifically related to the cascading effect of the cyber-attacks, namely the subversion of integrity, ways to measure the trustworthiness and integrity of data and systems (e.g. cybersecurity companies ratings for companies security performance, similar to credit scores) should be developed while realigning security research investments and priorities toward protecting trust, and work with allies to prepare for and reduce state attacks on integrity.[46]

6.5 Final Thoughts

As we are departing from the naïve first decades of the development of the cyberspace, cybersecurity presents itself as one of the great challenges of our times. The access to technology and to the know-how creates and augments vulnerabilities while at the same time our societies are more digitally connected than ever. Today, information, critical infrastructure, and trust in the integrity of political, economic, and social systems can all be targeted through a cyberattack and without the costs of an hefty war. Space systems are an especially vulnerable part of the cyberspace and have characteristics that make them particularly attractive to hackers.

Nevertheless, the national regulatory systems for cyber activities are still underdeveloped, perhaps this is one of the major reasons for the lack of proposals for the regulation of cyberspace at the international level. Even so, when the time comes to embark on this path, there are negotiation processes in other areas that provided important lessons and that allow not to make unnecessary mistakes.

Due to the close relationship between activities in space and cyberspace, which present similar challenges and share others, it is important to analyse and learn from the processes of regulating conventional military activities in outer space. One major lesson is that only a soft law instrument will succeed. Another lesson is that the proponents should not overestimate the aggregating power of a soft law instrument. The scope of the proposal should be duly outlined in proportion to the actors it wants to engage, namely if it addresses issues sensitive to national interests and strategies for space and cyberspace and the amount of issues one wants to address in a single legal instrument.

In this sense, the EU has an advantage to take the lead in the establishment of international norms of behaviour in cybersecurity. The European bloc inherently deals with this issues on an international level, which could make it a pioneer in the development of international standards for cybersecurity, allowing it to be one step ahead not only in the development, but more importantly, in the application of norms by several States. It is important for other countries to see soft law applied in this setting because its principles will be carefully analysed, and conclusions will be

[46]Neal A. Pollard, Adam Segal, Matthew G. Devost, 'Trust War: Dangerous Trends in Cyber Conflict' (War on the Rocks, 2018) <https://warontherocks.com/2018/01/trust-war-dangerous-trends-cyber-conflict/> accessed 2 April 2021.

taken on how these new norms are being applied. If the EU creates a working framework for cybersecurity, certain aspects might even be applied by other countries without the need to engage them, but merely by setting standards and practices that simply work.

João Falcão Serra holds a Master in European and International Law from the Nova School of Law of the Nova University Lisbon and is a member of the Space Law Investigation Center (SPARC) at the same University.

Lightning Source UK Ltd.
Milton Keynes UK
UKHW050109260822
407759UK00007B/12